Palgrave Macmillan Studies in Banking and Financial Institutions

Series Editor
Philip Molyneux, Bangor University, Bangor, UK

The Palgrave Macmillan Studies in Banking and Financial Institutions series is international in orientation and includes studies of banking systems in particular countries or regions as well as contemporary themes such as Islamic Banking, Financial Exclusion, Mergers and Acquisitions, Risk Management, and IT in Banking. The books focus on research and practice and include up to date and innovative studies that cover issues which impact banking systems globally.

Paul Wachtel · Giovanni Ferri ·
Ewa Miklaszewska
Editors

Creating Value and Improving Financial Performance

Inclusive Finance and the ESG Premium

palgrave
macmillan

Editors
Paul Wachtel
Stern School of Business
New York University
New York, NY, USA

Giovanni Ferri
Libera Università Maria SS. Assunta
Rome, Italy

Ewa Miklaszewska
Cracow University of Economics
Kraków, Poland

ISSN 2523-336X ISSN 2523-3378 (electronic)
Palgrave Macmillan Studies in Banking and Financial Institutions
ISBN 978-3-031-24875-7 ISBN 978-3-031-24876-4 (eBook)
https://doi.org/10.1007/978-3-031-24876-4

This Palgrave Macmillan imprint is published by the registered company Springer Nature Switzerland AG
The registered company address is: Gewerbestrasse 11, 6330 Cham, Switzerland

Introduction by the Editors

Back in 2007–2009 the financial sectors of developed countries, starting with the US, were caught by surprise and shaken by the subprime mortgage turmoil which then ballooned into the Global Financial Crisis. Fifteen years on, the multi-layered repair action has brought the financial sector to safer shores through two main internal evolutions: regulatory overhaul and improved business practice. A third evolution, external to the financial sector and yet crucial to restoring financial stability, has come through enhanced customers' financial education. However, technological progress and the transition to sustainable development have emerged as true game changers. Especially through digitalization and FinTech, the financial sector is deeply transforming in extraordinary ways, offering user-friendly approaches and innovative products. The second game changer raises an even bigger challenge—beyond innovating operational practice and products—requiring a revolution in mentality. It demands that finance transforms from self-referential to responsible. As society is asking for an acceleration of the transition to sustainable development, the financial sector is responding through the advance of Sustainable Finance, a whole new segment devoted to supporting the green transition, which is often referred to as ESG investing. In this acronym, which has rapidly become the compass of Sustainable Finance, E stands for Environment, S stands for Social and G stands for Governance. The idea is that the green transition caters to the environment but also to society and good governance may guarantee that companies should refrain from green/social washing.

This book collects papers analyzing how nowadays finance can create value and enhance performance. Specifically, the majority of the collected contributions address issues related to ESG or CSR topics and the role and impact of financial literacy. Environmental issues occupy a central position in almost any scientific field. In finance and banking, for a long time, they were at the outskirts of mainstream research, usually dealt with occasionally, as a part of bank corporate responsibility. Today, not only have non-financial business firms realized the importance of environmental protection, but also an immense awareness is seen among banks and financial firms, which is documented by the first half of the book.

Chapter 1: **The Market Reaction to Climate Risk: Evidence from the European Banking Industry,** by Francesca Battaglia, Francesco Busato and Simone Taddeo, utilizes data for the European banking industry to estimate the market reaction to climate risk and finds the reaction to be significant. In particular, examining a sample of the 45 major listed European banks over the period from 2014 to 2020, it explores the relationship between European banks' carbon emissions and stock returns before and after the introduction of the EU Sustainable Finance Action Plan in 2018. The results suggest that there is an indication that pro-environmental EU directives may influence market participants to shift away from firms that include high carbon intensive activities.

Chapter 2: **Dissecting the European ESG Premium vs the US: Is It All About Non-financial Reporting?** by Rocco Ciciretti, Ambrogio Dalò and Giovanni Ferri highlights the major gap between the ESG ratings of the top capitalized listed companies in the EU vs their analog counterparts in the US and attributes the EU premium to better non-financial disclosure in Europe.

In turn, Chapter 3: **Loan Origination and Monitoring Guidelines: How Do ESG Indicators Affect Firms' Probability of Default?** by Egidio Palmieri, Enrico Fioravante Geretto, and Maurizio Polato analyzes the impact of new EBA guidelines on credit risk management (2020 and 2021), which propose an innovative approach, orienting the credit analysis and management processes toward an anticipatory and proactive mode. A critical novelty is represented by the evaluation of Environment, Social and Governance (ESG) factors in determining firms' creditworthiness. The paper finds that ESG scores significantly contribute to reducing the Probability of Default of listed companies. Moreover, the size of this effect is found to be 3.5 times larger for EU companies than for their US homologues.

Finally, Chapter 4: **Using E from ESG in Systemic Risk Measurement** by Ewa Dziwok, Marta Anita Karaś and Michał Stachura presents a solution that allows using the ESG-scoring data in systemic risk analysis. They define environmental risk as the potential for adverse consequences for human or ecological systems that can arise from the impacts of environmental factors, including but not limited to climate change, as well as human responses to such factors. Since a sizable part of systemic risk comes from intermediaries' exposure to environmental risks, the paper stresses that the ESG data may be a source of information for systemic risk analysis and a fruitful approach seems to be augmenting systemic risk measurement with the E-factor derived from the ESG scores.

Chapter 5 follows the focus on non-financial factors shaping the financial markets' premium. **Corruption Disclosure in Banking: Insights from the Literature** by Pablo de Andrés, Salvatore Polizzi, Enzo Scannella and Nuria Suárez draws on the literature to assert that transparency and disclosure (both financial and non-financial) are pivotal to attenuate the adverse effects of information asymmetries in banking. Among the various dimensions of CSR disclosure, such as environmental disclosure, disclosure on human resources and community involvement, corruption disclosure deserves particular attention for building relationship and trust with the stakeholders and investors and generate more effective bank lending. The chapter describes the theoretical frameworks that can be adopted to analyze corruption disclosure in the banking sector.

The following two chapters (6 and 7) acknowledge the fact that financial literacy has been a core life skill, particularly for young people, allowing them not only to navigate modern society but also promoting financial resilience. In recent years, the financial markets are changing rapidly and individuals are increasingly called on to make more financial decisions than before; hence, financial education also plays a vital role in the policy agenda. Chapter 6: **Financial Competence and the Role of Non-cognitive Factors** by Elisa Bocchialini, Beatrice Ronchini and Francesca Torti gives interesting insights on how cognitive and non-cognitive components of financial education relate to each other. The paper surveys a sample of university students finding out a relatively high level of financial knowledge and a quite positive attitude toward finance and that correlation exists between attitudes toward finance and financial knowledge. This result has important implications for policy-makers: The attitude toward finance should be targeted to enhance financial knowledge among the young. On the other hand, Chapter 7: **Does Financial Literacy Progress Over Time? An Analysis of Three Surveys in Italy** by Paola Bongini, Doriana Cucinelli and Mariangela Zenga documents that in spite of major efforts through various programs, the level of financial literacy in Italy has declined in recent years. Consequently, a fundamental objective should be to improve the performance of financial education initiatives, which could comprise using also non-cognitive approaches. Their findings are important to define future financial education programs, as a current approach: "one-size-fits-all" cannot be successful considering socio-demographic differences highlighted by their analysis.

Chapter 8: **An Interdisciplinary Approach to Economic Texts: The "Considerazioni Finali" by the Governor of the Bank of Italy as a Case Study** by Paola Vezzani, Cristina Guardiano and Valentina Ligabue explores the content of the in-depth yearly speech, based on the Annual Report, by the Governor of the Bank of Italy, presenting the key-events concerning national and international economy. The paper analyzes the 75 CF speeches made from 1947 to 2021, investigating the relation between the content of these documents and their linguistic nature. It combines economics and linguistics approach to check whether their interaction provides insights which would not be reached through a monodisciplinary lens.

Finally, Chapter 9: **Drivers of Shareholder Value Creation in M&A: Event Study of the European Banking Sector in the Post-financial Crisis Era,** by Gimede Gigante, Mario Baldacchini and Andrea Cerri, employs an event study methodology to verify whether bank M&As created value for shareholders in post-sovereign crisis Europe. The creation of shareholder value involved in M&A transactions is one of the most discussed topics in the academic literature, however the 2008 financial crisis upset the traditional competitive logic in analyzing the ability of banks to generate value for shareholders, thus the chapter aims at identifying the new drivers guiding the reactions of market participants to announcement of an acquisition. The results indicate that the characteristics of a target financial institution are among the main determinants of a bidding company's abnormal returns.

To conclude, this book is another case in the post-2008 crisis discussion, asserting that the financial sector and especially banks can play an important role in promoting environmental sustainability and in contributing to value creation in the financial markets. For the latter, the main objective was to identify important dimension of value creation and enhancing financial performance. Thus, the book touches on very important problems—the role and areas of the financial sector activities related to environmental protection, governance issues and enhanced financial literacy. Those are crucial issues for the banking sector, financial markets and a large group of stakeholders, including shareholders and bank clients. If we can adequately address these issues, this may have an important impact toward creation of a sustainable financial and banking market.

Paul Wachtel
Giovanni Ferri
Ewa Miklaszewska

Contents

Notes on Contributors

Pablo de Andrés is Professor of Finance and Head of the Finance Unit at the Autonomous University of Madrid. He is also Research member of the European Corporate Governance Institute (ECGI) and Honorary Fellow of the Hanken School of Economics. He has focused his research on the analysis of financial decisions, real options, and corporate governance. His interest in corporate governance started early in his career, as he was awarded the European Investment Bank Prize for young researchers (1997) for his work "Financial system models, corporate governance and capital investment in OECD countries: Some stylized facts." Since then, he has published his research in finance and business high-impact journals. In the last years, he is specially interested in the interplay between corporate governance, finance, and technology. He set up a lab with the start-up Collisio Technologies, where a multidisciplinary team researches on the financing strategies through digital channels and assets.

Mario Baldacchini is currently Investment Analyst in the alternative asset management team at Equita in Milan. Previously, he gained experience in the finance industry at Menara Capital, an investment boutique based in Paris, Mediobanca and Intesa Sanpaolo in Milan. He holds a MSc degree in Management with major in corporate finance from Bocconi University.

Francesca Battaglia is Associate Professor of Banking and Finance at University of Naples "Parthenope," Department of Management Studies and Quantitative Methods. She also serves as a member of the teaching board of the Doctorate "Governance, Management and Economics," within the Department of Economics and Management. Her research has focused on risk management, securitization, corporate governance, crowdfunding, and systemic risk of financial intermediaries. She published papers in a number of academic referred journals, and she serves as referee for European Journal of Finance, International Review of Financial Analysis, Journal of International Markets Institutions and Money, European Financial Management and Corporate Governance: An International Review. She is also a member of the Scientific Committee of Arcane Finance & Development Publishing Series. She was awarded "Highly Commended Paper award 2015" within "Emerald Literati Network."

Elisa Bocchialini is Assistant Professor in Financial Intermediaries at the Department of Economics and Management of the University of Florence, Italy, where she teaches "Bank Management" and "Financial Institutions." Financial literacy is one of her research interests.

Paola Bongini is Full Professor of Banking and Finance at the University of Milano-Bicocca (Unimib, Italy), where she teaches Investments and Financial Institutions Management. She is also Director of the Ph.D. Program in Business for Society at Unimib. Her research interests are in empirical banking and financial regulation. Her most recent interest includes financial literacy and financial education, serving as a member of the Scientific Committee of the Italian Foundation for Financial Education (Feduf). Her works have been published in the Journal of Banking and Finance, Sustainability, Small Business Economics, among others. She has been involved in several national and international research projects and in relevant outreach activities in both the financial and nonprofit sector. Outside academia, she seats on the board of directors of an Italian significant bank.

Francesco Busato is Full Professor of Economics at the University of Naples "Parthenope," Department of Economic and Legal Studies, Italy. He served as Assistant Professor at University of Aarhus Denmark, University of Rome III, as Associate Professor and next Full Professor at the University of Naples Parthenope. He is the Head of Ph.D. Program of Economics, Statistics, and Sustainability. He also serves as Deputy Director of the Department of Economic and Legal Studies at the University of Naples Parthenope. He teaches classes in topics such as finance, economics, and bachelor, master, and Ph.D. level. He has published several papers in a number of peer-reviewed international journals in finance, economics (dynamic models), and sustainability.

Andrea Cerri is Academic Fellow of the Department of Finance at Università Bocconi and at SDA Bocconi. He currently holds the course of Finance Lab at Università Bocconi. He also teaches International Financial Markets and then Venture Capital and managerial entrepreneurship for IES Abroad, Milan Center. He carried out post-doctoral research activities at the Faculty of Economics of Università Cattolica in Milan. He is author of national and international articles and publications on financial topics with several editors like Palgrave Macmillan Studies in Banking and Financial Institutions and Bocconi University Press (BUP). As a professional, he is enrolled as Chartered Professional Accountant, Statutory Auditor, and he is enrolled ad Technical Consultant of the Court of Milan. As a professional, he serves in some boards of statutory auditors; he supports firms and financial institutions in litigations, operates as firms' technical consultant, and provides services like firms' valuation or business plan implementation or restructuring plans.

Rocco Ciciretti is Associate Professor of Economic Policy at Department of Economics and Finance, University of Rome Tor Vergata, where he currently teaches two graduate courses: financial market models and applied finance. During the 2016–2018, he coordinated the graduate program in financial markets and financial intermediation. He is board member of COMETA pension fund, and President in the exam committee II of the Organismo di vigilanza e tenuta dell'albo unico dei Consulenti Finanziari (OCF). He is Research Fellow at RCEA—Rimini, and at CEIS—University of Rome Tor Vergata where he is also Assistant Editor for the CEIS Research papers series. He was Research Fellow at SIRP and at EPRU—University of Leicester School of Management, and Visiting Professor at New York University, Fordham University, Rensselaer Polytechnic Institute, SOM Research Institute—University of Groningen. He has published and is a referee for several international journals. He served as financial economist and assistant policy advisor in the Finance Team at the Research Department of the Federal Reserve Bank of Atlanta. Prior to joining the FED, he was Research Associate in Finance at the Lally School of Management and Technology—Rensselaer Polytechnic Institute in Troy, NY.

Doriana Cucinelli is Assistant Professor in Banking at the University of Parma, Italy. She is also Associate Researcher since March 2020 and member of Panel Experts since 2008 at Euro-Mediterranean Economists Association (EMEA). Her main research interests are risk management in banking, bank business models, and financial literacy. She is author of national and international publications and has presented at many national and international conferences. Her works have been published in the Journal of Banking and Finance, Journal of Financial Stability, British Journal of Management, among others. She holds a Ph.D. in Banking and Finance at University of Roma Tor Vergata and a Laurea cum laude in Finance and Risk Management at the University of Parma. She was visiting fellow at Bayes Business School (formerly Cass) in 2017. Outside academia, she seats on the board of directors of an Italian cooperative bank.

Ambrogio Dalò is a tenured Assistant Professor of Finance at the Department of Economics, Econometrics, and Finance—University of Groningen (The Netherlands), where he currently teaches both at the graduate and at undergraduate levels. He also supervises both Bachelor and Master theses. He serves as Coordinator of the Bachelor Thesis for the Business and Economics program. He holds a Ph.D. in Economics and Finance (European Label) and a M.Sc. in Quantitative Finance from the University of Rome "Tor Vergata." He earned a bachelor's degree in Management from the University of Bari "Aldo Moro." His current research interests are Empirical Asset Pricing, Responsible Investing, and Mutual Funds.

Ewa Dziwok is Professor at the University of Economics in Katowice, Faculty of Finance, and Head of the Department of Applied Mathematics. She is also Director of Master studies Quantitative Asset and Risk Management ARIMA. Her research focuses on the applications of quantitative methods in finance and recently on the greening financial system processes. She has authored many publications and participated in various international research projects, working in cooperation with European universities and central banks.

Giovanni Ferri is Full Professor of economics at the Department of Law, Economics, Politics, & Modern Languages at LUMSA University in Rome where, from 2014 to 2018, he was Deputy Rector for Academic Affairs and chaired the Quality Assurance and Library Committees. He also co-founded and chairs the Center for Relationship Banking & Economics to foster research for a better understanding of relational goods to promote society wellbeing. He founded and chaired the Master in Management of Sustainable Development Goals and five other executive masters in sustainable management. He previously served at the University of Bari, World Bank, Banca d'Italia. He visited: ADB Institute, HKMA, NBER, Princeton University, Tokyo University. He consulted for EU Commission/Parliament, Italy's Treasury and partook in EBA's Banking Stakeholder Group. He is editor-in-chief of Economic Notes and led (or consulted for) policy/research projects in many countries, including Albania, China, Egypt, Indonesia, Korea, Lebanon, Malaysia, Philippines, Serbia, Thailand, Tunisia, Turkey, and Vietnam. To support the sustainable transition by providing innovative consulting and knowledge building to both public and private sector organizations through an ethically sound business model, he has co-founded (and is scientific chair) of Sustinentes. He has published extensively on/for several world recognized journals/publishers on various fields: financial shocks, banking structure, emotional intelligence and happiness, rating agencies, corporate governance, migration, firm internationalization, Chinese economy, family firms, inequality, and sustainability.

Enrico Fioravante Geretto is Associate Professor at the University of Udine, Department of economics and Statistics. In the past, he was Lecturer of Banking Operations Technique, Insurance Operations Technique, Financial Instruments Technique, Asset Management Economics, Intermediaries, and Financial Markets. Currently, he is Professor of Economics and Management of the Bank (master's degree in Banking and Finance) and Economics of Financial Intermediaries.

Gimede Gigante Ph.D., Course Director of Principles of International Finance at Bocconi and Harvard Business Review advisory council member, has served as Academic Director of the Bocconi Summer School since 2019. Deputy Director of the Master of Science in Finance, he is also Deputy Director of the Bachelor Degree Program in Economics and Finance at Bocconi University, where he holds the academic position of Lecturer in the Finance Department. He holds ITP qualification (International Teachers' Program) from SDA Bocconi. He has held visiting positions at the Finance Department of Columbia Business School and at the Salomon Brothers Center (Stern School of Business, NYU). He is a fellow of the Research Unit of "Investment Banking & Structured Finance" at the Baffi-Carefin Center (Centre For Applied Research on International Markets, Banking, Finance, and Regulation). He has been teaching Financial Systems, International Finance, and Investment Banking courses as a faculty member of the Master Program in International Management and the Master of Finance at Bocconi University in Milan. Professor of Investment Banking at SDA Bocconi in Mumbai (India) since 2014, he has been teaching Finance Lab & Private Equity at the Summer School of Università Bocconi (since 2015), and Corporate Finance at Fudan University (Shanghai-China) as part of the Fudan-Bocconi Double Degree in International Management Program (DDIM) (since 2016). In 2022, his courses included Private Equity at SDA Bocconi's MBA Master Program with Seoul National University (SNU). Certified Public Accountant and Professional Auditor, his main areas of research are international finance, financial markets, corporate valuation, investment banking, fintech, and private equity. He has published a variety of papers on banking and acts as a consultant and board member to several financial and non-financial institutions.

Cristina Guardiano is Full Professor of Linguistics at the Università di Modena e Reggio Emilia (Italy). She specialized in classics and linguistics at the Università di Pisa, where she also got her Ph.D. in historical syntax, with a dissertation about the internal structure of DPs in Ancient Greek. She is active in research about formal and quantitative linguistics, crosslinguistic comparison, language change, phylogenetic reconstruction, and gene-language comparison. She is specialized in the parametric analysis of nominal phrases and in the comparative study of diachronic and dialectal syntactic variation. She is a Fulbright *alumna* (2013–2014, Department of Linguistics, UCLA), has been chief Research Associate of the ERC advanced grant (ERC-2011-AdG) n° 295733 LANGELIN (2012–2018, PI: Giuseppe Longobardi, University of York; Co-I Guido Barbujani, Università di Ferrara) and is currently a member of the SSWL (Syntactic Structures of the World's Languages) research team: She is active as a property author (with Hilda Koopman), language expert, and administrator. She is also a member of the TerraLing board. She currently coordinates the Reggio Emilia research unit in the MIUR PRIN project "Models of language variation and change: new evidence from language contact" (2019–2023), and she is a member of the project ITHACA .

Marta Anita Karaś is Assistant Professor at the Department of Financial Investments and Risk Management at Wrocław University of Economics and Business, Faculty of Economics and Finance. During her master's and Ph.D. studies, she graduated from international programs at the London School of Economics and Political Science, Stony Brook University in New York, and Universitat Pompeu Fabra in Barcelona. Her research interests focus on financial markets, monetary policy, and systemic risk analysis. Over the last years, she has carried out four research projects focused on the application of existing and new systemic risk measurement methods to the CEE region, both individually and in cooperation with colleagues from various European Universities and the National Bank of Poland.

Valentina Ligabue is a graduate of Modena and Reggio Emilia University with a master's degree in Business Management and Communication. She previously graduated with a degree in foreign languages having studied in both the UK and USA. She is currently working in the International Finance Department of an accounting and legal firm in Reggio Emilia, focusing on strategic planning at the international level aimed at optimizing the overall taxation processes of corporations.

Ewa Miklaszewska is Professor of Banking and Finance at Cracow University of Economics, where she heads the Department of Banking and Global Financial System at the Institute of Finance. She is also Lecturer at the Jagiellonian University in Cracow, Poland, and in the Summer School of Radboud University in Nijmegen. She was a Visiting Professor at various universities, including NYU Stern School of Business. She is also a member of the Managing Committee of The European Association of University Teachers in Banking and Finance—the Wolpertinger Group. Her research focuses on the impact of regulatory policies on bank performance and on factors influencing bank strategies and business models.

Egidio Palmieri Ph.D., is Researcher at the University of Udine, Department of Economics and Statistics. He deals with risk management in banks and the implications of ESG issues within the financial system.

Maurizio Polato is Full Professor of Banking and Finance at Udine University, Italy. His research fields mainly relate to the securities and exchange industry and bank performance. He is the author of various publications on the topic, which address issues related to the industry's structure, performance, and value measurement for securities trading and regulation.

Salvatore Polizzi, Ph.D., is Assistant Professor in Banking and Finance at the Department of Economics, Business and Statistics, University of Palermo (Italy), where he teaches Banking and Financial Intermediation. His research interests include banking, risk management, and disclosure. During his academic career, he worked for the European Central Bank and Banca d'Italia. He held Visiting Researcher position at the Bangor Business School, Bangor University, UK. He is member of the European Association of University Teachers in Banking and Finance (Wolpertinger club), of the Italian Association of University Teachers in Banking and Finance (ADEIMF), and of the Italian Academy of Management (AIDEA). His publications appear in reputable academic journals in the field of banking and finance, including the *Journal of International Money and Finance*, the *Review of Quantitative Finance and Accounting*, and *European Financial Management*, among others. He is also author of a monograph entitled "Risk Disclosure in the European Banking Industry," published by Springer in 2022.

Beatrice Ronchini is Associate Professor of Financial Markets and Institutions at the Department of Economics and Management of the University of Parma, Italy, where she teaches "Economics of Financial Intermediation" and "E-Banking."

Enzo Scannella is an M.B.A., Ph.D., Professor of Banking and Finance at the Department of Economics, Business, and Statistics, University of Palermo (Italy), where he teaches Banking, Economics and Management of Financial Institutions, Risk Management. His main research interests are focused on banking, financial regulation, risk management, risk disclosure, and financial innovation. He has held visiting positions in foreign universities, including Harvard Business School (Boston, USA), Universidad Autónoma de Madrid (Madrid, Spain), and Carleton University (Ottawa, Canada). His publications appear in leading nationally and internationally recognized academic journals. He has also contributed to a number of books. He is a member of various scientific associations, university research groups, editorial boards, and scientific committees for conferences. He is a member of the European Association of University Teachers in Banking and Finance (Wolpertinger), the Italian Academy of

Management (AIDEA), and the Italian Association of University Teachers in Banking and Finance (ADEIMF).

Michał Stachura is Assistant Professor at the Department of Economics and Finance of the Faculty of Law and Social Sciences of the Jan Kochanowski University in Kielce, Poland. The area of his research interests includes the theory and applications of statistics and econometrics in finance, health economics, and broadly understood social sciences. He has been involved in several interdisciplinary research projects in cooperation with various universities and national research institutes, where he applied his econometrics and statistics expertise in geological, legal, and systemic risk studies.

Nuria Suárez, Ph.D. (Extraordinary Doctoral Award), University of Oviedo (2010), is member of the Finance Department at the Universidad Autónoma de Madrid since 2017.Her research interests focus on the banking and corporate finance field.She has published articles in recognized international finance journals, such as Journal of Banking and Finance, Journal of Corporate Finance, Journal of Financial Stability, Journal of International Financial Markets, Institutions & Money, Journal of International Money and Finance, International Review of Economics and Finance, or International Review of Law and Economics, among others. She is also author and coauthor of book chapters published by Springer and Nova Science Publishers. She usually attends to the main international and national research conferences in banking and finance and has been member of several research projects during the last ten years. She is currently the main researcher of a research project granted by the Comunidad de Madrid. She has been Visiting Professor at the University of Palermo (2018), in Italy. During her doctoral studies, she visited the Business Department of the Carlos III University of Madrid (2010).

Simone Taddeo is a Ph.D. student in Economics, Statistics, and Sustainability at the University of Naples "Parthenope," Department of Economic and Legal Studies, Italy. He has been a Visiting Bachelor Scholar at the University of Groningen (Netherlands) and a Visiting Master scholar at the University of New Jersey Institute of Technology, USA. He has been a Ph.D. Visiting Research Scholar at the Leuphana University of Lüneburg, Germany, and at the University of Waterloo, Canada; he has also served as university tutor in Institutions of Economics and Financial Mathematics at the Telematic University of Studies Italian

University Line. His research cover topics on ESG Investments and Sustainable Finance, Corporate Finance, and Financial Econometrics.

Francesca Torti got a Ph.D. in Statistics at the University of Milano-Bicocca, and she is chartered Accountant and Auditor. She conducts research in statistics, with particular focus on robust statistical methods. She currently works at the Joint Research Center of the European Commission.

Paola Vezzani is Full Professor of Banking and Finance at the Department of Communication and Economics of the University of Modena and Reggio Emilia, Italy, where she was previously the Dean of the Faculty and the Head of the Department. She graduated from the University of Modena and obtained a Ph.D. in Capital Markets and Financial Management from the University of Bergamo. She was a visiting Ph.D. student at the Graduate School of Business Administration in the Finance Department of New York University. She is a member of CEFIN (Center for Studies in Banking and Finance), AIDEA (Italian Academy of Business Administration and Management), ADEIMF (Italian Association of University Teachers of Financial Intermediaries, Markets and Finance), and the European Association of University Teachers in Banking and Finance. She is a member of the scientific committee of Banking and Financial Diploma of the Italian Banking Association (ABI). Her research topics include the diversification strategies of banking groups, organizational and business models, bank performance analysis, intellectual capital, customer complaints, non-financial reporting, banks' remuneration policies, corporate governance, and the Nobel Prize in Economic Sciences.

Paul Wachtel Professor Emeritus of Economics, was a member of the faculty of the New York University Stern School of Business from 1972 to 2021. He served as the chairperson of the Economics Department, the Vice Dean for Program Development, Academic Director of the BS in Business and Political Economy, and the chairperson of the University Faculty Council. His primary areas of research are monetary policy, central banking, and financial sector reform in economies in transition. His research has been published in numerous journals, including, most recently, *Brookings Papers on Economic Activity*, Journal of Money Credit and Banking, Journal of Finance, Economic Inquiry, and Journal of Banking and Finance. He was the co-editor of Comparative Economic

Studies and is Associate Editor of the Journal of Banking and Finance. He has served as the program chair for the Dubrovnik Economic Conferences for over 20 years and was previously Research Associate at the National Bureau of Economic Research, and a senior economic advisor to the East West Institute. He is the chair of the Academic Board of the International School for Economics in Tbilisi and received a DOCTOR HONORIS CAUSA from the Bucharest University of Economic Studies in 2016. He is also very actively influencing public policy, by providing opinions and comments in influential journals, such as Business Insider, American Banker, Forbs, and Financial Times.

Mariangela Zenga is currently Associate Professor in Social Statistics at the University of Milano-Bicocca, Italy. Her main research interests are in statistical methods for the evaluation of healthcare systems, labor market, tourism services, and educational systems (paying attention to the gender differences).

List of Figures

List of Tables

CHAPTER 1

The Market Reaction to Climate Risk: Evidence from the European Banking Industry

Francesca Battaglia, Francesco Busato, and Simone Taddeo

1.1 Introduction

Nowadays, discussion on climate change is a widespread issue in the world debate. The mean global temperature of the Earth has seen an increase of 0.87 °C since 1900 according to Intergovernmental Panel on Climate Change (IPCC, 2014). At this rate, if detected emission levels since 1950

F. Battaglia
Department of Management Studies and Quantitative Methods, University of Naples Parthenope, Naples, Italy
e-mail: francesca.battaglia@uniparthenope.it

F. Busato · S. Taddeo (✉)
Department of Economic and Legal Studies, University of Naples Parthenope, Naples, Italy
e-mail: simone.taddeo001@studenti.uniparthenope.it

F. Busato
e-mail: francesco.busato@uniparthenope.it

P. Wachtel et al. (eds.), *Creating Value and Improving Financial Performance*, Palgrave Macmillan Studies in Banking and Financial Institutions, https://doi.org/10.1007/978-3-031-24876-4_1

continue to rise, global warming is likely to reach 1.5 °C above pre-industrial levels between 2030 and 2052 (IPCC, 2018). The progressive increase in global warming is destined to cause unstoppable catastrophes, provoking a significant impact not only for humankind, but also for the ecosystem and natural resources. Scientific articles by the IPCC reveal that one of the causes of global temperature rise is the continuous increasing concentration of greenhouse gas emission in the atmosphere (IPCC, 2014).

A rise of the greenhouse gases concentrations in the air produces a significant climate forcing, or warming effect. Over the period that goes from 1990 to 2019, the global warming effect provoked by human activities' greenhouse gases increased by almost 45% (Environmental Protection Agency, 2021). Therefore, it is plausible to believe that one of the reasons that the planet heats up is largely due to anthropogenic activities (human activities), which is considered the biggest contributor to climate change (U.S. Global Change Research Program, 2021). In order to reduce the global warming effect, one of the main recommendations that scientists suggest is to lower greenhouse gas emissions, encouraging the transition to a low-carbon economy (IPCC, 2014).

A first step toward the abatement of gas emissions was made by the Paris Agreement in 2015, an international treaty where most of the countries representing 97% of worldwide greenhouse emissions, agreed to keep global warming below 2 °C, preferably at 1.5 °C, compared to pre-industrial levels. The Paris Agreement, which represents a real milestone for combating climate change, raised awareness among policymakers, academics, financial institutions and companies regarding the variability of weather's temperatures as a future challenge and concrete threat in the next decades (ESRB, 2016). The Paris treaty can be considered the first climate deal that has contributed to rethinking a new way of doing business by favoring the transition from an economy with a high greenhouse gas emission to a low-fossil-fuel-economy (LFFE) or commonly called low-carbon economy.

One of the three long-term goal commitment of the agreement, indeed, was to "making finance flows consistent with a pathway towards low greenhouse gas emissions and climate-resilient development" (Paris Agreement, Article 2.1c).[1] In this context, the European Union has taken

[1] https://unfccc.int/sites/default/files/english_paris_agreement.pdf.

giant steps in reducing its greenhouse gas (GHG) emissions. The latest statistics show that GHG emissions in Europe have decreased sharply in the last years, reaching 24% below 1990 levels, which is expected to be 31% in 2020 (European Environment Agency, 2021). This is attributable not only to the transition of the use of fossil fuels to a clean energy source (such as renewable resources), which has led to structural changes in European economies, but also to the implementation of EU and national policies and regulations. On 7 March 2018, indeed, the European Commission launched one of the most important action plans for financing sustainable growth, facilitating the transition to a low-carbon economy by increasing investments in green projects and promoting a new financial sustainability strategy in the long-term.

The policy is well-known under the name of Sustainable Finance Action Plan, whose ultimate goal is to shape the financial system in a way to support the sustainable transition. The Action Plan recommends three key objectives to be taken at European level. The first purpose is to redirect cash and capital flows toward sustainable investments shifting away from those activities and sectors that make intensive use of fossil fuels that encourage the global warming issue. The second goal is to manage financial risks deriving from climate change, resource depletion and environmental degradation. The third aim is to enhance the transparency and long-termism in each financial activity as to realize sustainable and inclusive growth. These three objectives are divided into ten actions which include initiatives on various fronts with the aim of involving all the players in the financial system in reducing information asymmetries related to climate risks, thus improving the allocation of capital to sustainable investments. In detail, the scope of the action plan encourages to better classify economic activities along with an appropriate EU sustainable taxonomy, clarifying to all market participants, such as asset managers, pension funds, and European banks their responsibilities regarding sustainability. This allows the possibility to assess the feasibility of including the risks associated with the climate and other environmental factors in the risk management policies.[2]

[2] European Commission (2018), communication from the Commission to the European Parliament, the European Council, the Council, the European Central Bank, the European Economic and Social Committee, and the Committee of the Regions Action Plan: Financing Sustainable Growth. https://eur-lex.europa.eu/legal-content/EN/TXT/PDF/?uri=CELEX:52018DC0097&from=EN.

In this view, financial institutions are not exempt in increasing their transparency only on the integrated sustainability in investment decisions levels, but also in reference to their own activities. Indeed, banks contribute to GHG emissions in two ways: indirectly and directly.

On the first hand, banks help to raise global warming through the financing of companies' projects and the implementation of firms' business plans which, in turn, boost GHG emissions. Even nowadays, banks continue to play a major role as principal financiers of the most polluting sectors, such as coal, oil, and gas industry, delaying the transition from an economy with a high use of coal and fossil fuel-based to an efficient renewable and sustainable-based economy. However, at the same time, financial institutions play an important role in managing climate change, by helping the overall financial system to move toward a low-carbon economy (Bank of England, 2021; European Central Bank, 2021; FED, 2021).

On the other hand, banks, such as all companies, generate greenhouse gases (GHG) directly through their activities. According to the European Central Bank report on institutions' climate-related and environmental risk disclosures (2020),[3] only 8% of all financial institutions report the percentage of carbon-related assets in each portfolio, 14% the carbon footprint of one or more portfolios where the main metrics reported are referred to Scope 1, Scope 2, and Scope 3 carbon emissions.

Following the Greenhouse Gas Protocol standards, which determines the criteria for measuring carbon firm issue, carbon footprint for each entity could be grouped into three macro-classes, defined by the GHG Protocol Scope 1, Scope 2, and Scope 3. This label categorization is necessary to show which emission sources are under the direct control of the company, and such under the control of other third-parties' organizations, in order to distinguish between direct and indirect emission sources.

From the need to mitigate climate change, carbon footprints and disclosure metrics are thus considered important tools not only for the safeguard of the environment, but also for investors in their investment decision strategies (Krueger et al., 2020). Indeed, there is a growing

[3] ECB (2020), available at https://www.bankingsupervision.europa.eu/ecb/pub/pdf/ssm.ecbreportinstitutionsclimaterelatedenvironmentalriskdisclosures202011~e8e2ad20f6.en.pdf.

strand of literature that certifies a strong correlation between the intensity of carbon emission and financial performances. Some authors claim for positive relationship, meanwhile others state the opposite (Boiral et al., 2012; Gallego-Álvarez et al., 2015; Hatakeda et al., 2012; Ziegler et al., 2007).

In this context, it is, therefore, essential that carbon emissions of firms are constantly monitored by investors in order to obtain superior returns (Siddique et al., 2021). At the same time, investors should consider in their investment decision process also the effects of environmental regulation. According to Krueger et al. (2020), a great percentage of financial institutions declare that climate risks related to new regulations are already in place. Therefore, depending on the stringency of the regulation, investors would demand superior returns for those firms that have higher environmental regulation's risk (Testa et al., 2011).

By applying a panel data analysis over 45 listed European banks, spanning from the period that goes from 2014 to 2020, we find that before the introduction of the EU Sustainable Finance Action Plan, investors required higher compensation for those financial institutions that included carbon-intensive activities, as a hedge against climate risk. After the launching of the European sustainable guideline, the correlation between banks' carbon emissions and returns dropped, leading investors to initiate the so-called portfolio decarbonization process. Our work is organized as follows. Section 1.2 reviews the academic literature about carbon premium and specifies our research questions. Section 1.3 describes our sample and data sources, introducing the variables used and the specification model. Section 1.4 comments our main results meanwhile Sect. 1.5 concludes.

1.2 Research Hypothesis and Related Literature

The problem of climate change leaves neither investors nor companies indifferent. From this point of view, investors can no longer stand by and ignore the impact that global climate change has on their portfolios. The emerging economic literature offers both theoretical and empirical proofs that financial actors should take carbon-transition risk into account in their investment decision process (Ilhan et al., 2021; Krueger et al., 2020). In this view, investors will seek a greater economic return, depending on carbon risks and opportunities. The literature about the relationship between carbon emissions and corporate financial performance is still scant, and the empirical evidence offers different and ambiguous results. For example, some authors claim for a "carbon premium" (Karydas & Xepapadeas, 2019; Pindyck, 2013; Tol, 2008). Bolton and Kacperczyk (2020) discovered that the carbon emission impacts positively and significantly the US firms' stock returns in both direct and indirect emissions. Again, Bolton and Kacperczyk (2021b) estimated a market-based carbon premium in a cross section where they take in consideration almost 80 countries with 14,400 firms analyzed. They found out that greater returns are related with higher levels and growth rates of carbon emissions mostly in all sectors suggesting the so-called carbon premium, i.e., higher stock price returns are retained to be for all firms that are considered carbon inefficient.

In other words, investors would require higher compensation for those firms that are highly exposed to climate change risks. Along the same idea Oestreich and Tsiakas (2015) provided an empirical investigation about the consequence of the European Union's Emissions Trading Scheme on German stock returns. They noticed that those firms in charge of free carbon emission allowances presented higher returns in relation to firms that did not. As a result, they found a presence of "carbon premium" mainly clarified by greater cash flows because of the free carbon emission allowances. Weitzman (2009) and Litterman (2013) claim that investors demand higher returns to polluting companies as compensation for climate risk-taking, as a result of climate change. Lastly, Monasterolo and De Angelis (2020) found out that after the Paris Agreement signed in 2015, investors require higher compensation premia for those assets which are involved with carbon-intensive sectors.

However, there are new studies that find that the nature of the relationship between carbon emissions and financial performance is not always positive (Alvarez, 2012; Gallego-Álvarez et al., 2014). Indeed, a different strand of literature suggests a negative relationship between firms' carbon emission and stock returns (Delmas et al., 2015; Galema et al., 2008; Heinkel et al., 2001). For example, Matsumura et al. (2014) show that firms' value is linked with levels emission in negative way, i.e., displaying a decrease in terms of value for each higher level of carbon released. Again, Busch and Hoffmann (2011) report an inverse relationship between carbon emission intensity and different financial performance measures, such as Tobin's Q. Consistent with this result is the work of Aggarwal and Dow (2013), where claim for a negative correlation.

Other studies display that stock market downweighs firms that act badly in environmental terms in favor of those which perform well (Bernardini et al., 2021; Ferrell et al., 2016; Trinks et al., 2020).

Regardless of the nature of the relationship, the debate about this relationship has been stimulated mainly by the tightening of environmental regulations, increasing investors' environmental perception and awareness. A study conducted by Hsu et al. (2020) demonstrated that firms which pollute the most are the more vulnerable to environmental regulation' risks. This is in line with the vision of Bolton and Kacperczyk (2020), where it is stated that firms that release disproportionately levels of CO2 emissions are more exposed to regulatory interventions. According to Chapple et al. (2013) instead, good environmental performance eludes financial risks arisen by direct and indirect costs, such as increased regulatory intervention. Therefore, environmental regulatory and directive stringency affect stock prices and returns (Hong & Kacperczyk, 2009; Pastor & Veronesi, 2012).

Examining a sample of the 45 major listed European-zone banks over the period from 2014 to 2020, the aim of this study is to explore the relationship between European banks' carbon emissions and stock returns before and after the introduction of the EU Sustainable Finance Action Plan in 2018.

This work contributes to the strand of literature on the topic in several ways. Firstly, despite a number of studies have analyzed the impact of carbon emissions on non-financial firms' returns, ours is the first attempt that takes into consideration only the European banking industry. For this reason, this study is unique in its kind since there is no regression analysis in the literature that looks at the European banking sector. Secondly, this study contributes to the literature that analyses the effect of the EU Sustainable Finance Action Plan on the European banking industry emissions. Indeed, in this chapter, we investigate the impact of climate change-related policies on European banks' emissions. Thirdly, this work focuses on evaluating a bank's exposure to climate risks and on examining the investors' reaction. What is found out is that stock market participants take carbon emissions in consideration in their investment decisions, in view of the so-called portfolio decarbonization process, aimed at reducing GHG intensity by entering low-carbon activities stocks. Overall, the aim of this analysis enlightens the following two testable hypotheses:

Hypothesis (H1) Banks which include high-carbon activities have higher returns prior to the introduction of the EU Sustainable Finance Action Plan in 2018.

Starting from this point, our next consideration comes to light which brings us to specify the second hypothesis to be tested in our work:

Hypothesis (H2) After the introduction of the EU Sustainable Finance Action Plan in 2018, high-carbon-intensive banks have lower returns as a result of the European environmental guideline.

1.3 Sample Data and Methods

The list of sample entities refers to 45 listed European banks with a market capitalization higher than 1 billion of Euros with a time horizon that goes from 2014 to 2020. In total, European banks belong to 17 different countries. These are: Austria, Belgium, Czech Republic, Denmark, Finland, France, Germany, Greece, Holland, Hungary, Ireland, Italy, Norway, Portugal, Spain, Sweden, and lastly, UK. Officially, UK left Europe on 31 January 2020, but nevertheless, both sides agreed to keep many things the same until 31 December 2020, to allow enough time to agree to the terms of a new trade deal. In addition to that, the UK government reiterated its intention to maintain an equivalent regime to manage cross-border finance activities. So, the UK sustainable goal matches the ambition of the European Union's sustainable finance action plan through a series of package decisions being granted before the end of the transition period and beyond, in maintaining dialogue with the EU. The same goes for Norway, which, because of the Agreement on the European Economic Area (EEA), ensures the participation in the EU internal market. Table 1.1 reports the list of the European banks taken in consideration in our analysis.

Consistently with our purposes, our dependent variable is the weekly annualized stock bank return. As explanatory variables, we collect different selection of indicators that measures the strength of different business activities extrapolated from the bank balance sheet and income statement information from Thomson Reuters Eikon. Yearly data are acquired.

Table 1.1 European Banks sample

Bank name	*Country*
Aareal Bank AG	Germany
ABN AMRO	Holland
AIB Group plc	Ireland
Alpha Services and Holdings SA	Greece
Banca Monte dei Paschi di Siena SpA	Italy
Banco Bpm	Italy
Banco Comercial Portugues SA	Portugal
Banco de Sabadell	Spain
Bankinter	Spain
Barclays PLC	UK
BAWAG Group AG	Austria
BBVA	Spain
BNP Paribas	France
Bper Banca	Italy
Caixabank	Spain
Close Brothers Group PLC	UK
Commerzbank	Germany
Credit Agricole	France
Danske Bank A/S	Denmark
Deutsche Bank	Germany
Erste Bank	Austria
Eurobank Ergasias Services and Holdings SA	Greece
HSBC Holdings PLC	UK
ING Groep	Holland
Intesa	Italy
Jyske Bank A/S	Denmark
KBC Groep NV	Belgium
Komercni Banka as	Czech
Lloyds Banking Group PLC	UK
Mediobanca	Italy
National Bank of Greece SA	Greece
Natwest Group PLC	UK
Nordea Bank Abp	Finland
OTP Bank Nyrt	Hungary
Piraeus Financial Holdings SA	Greece
Raiffeisen Bank	Austria
Santander	Spain
Skandinaviska Enskilda Banken AB	Sweden
Société Générale	France
Sparebank 1 SR Bank ASA	Norway

(continued)

Table 1.1 (continued)

Bank name	*Country*
Standard Chartered PLC	UK
Svenska Handelsbanken AB	Sweden
Swedbank AB	Sweden
Sydbank A/S	Denmark
UniCredit	Italy

The table reports the list of the 45 European banks used in our sample. The sample period is 2014–2020

We include several bank-specific control variables to account for the bank performance. Following the work of Bolton and Kacperczyk (2021a), we consider Total Assets, Book to Market, Debt on Equity as proxies of size, valuation, and leverage ratio in our main regression. In addition to that, we take in account other financial metrics, such as Profit Margin, Loan to Deposit, and Capital Adequacy Ratio to account for profitability, liquidity, and solvency (European Banking Authority, 2021).[4]

In relation to our key independent variables, we include determinants of carbon emissions (Bolton & Kacperczyk, 2021a). Thomson Reuters Eikon database provides unique annual information for each bank-level carbon emission and related Greenhouse Gas emission (GHGs) for all over the sample period analyzed. Indeed, the databank accounts all three scopes of carbon emissions in units of tons of CO_2 emitted yearly.

The Scope 1 is linked to the direct GHG emissions from installations within the borders of the organization due to the use of fossil fuels and the emission of any greenhouse gas into the atmosphere. Direct emissions are, for example, the emissions deriving from the combustion of fossil fuels in heating systems; emissions due to the consumption of fuel for company vehicles; leaks of fluorinated greenhouse gases from air conditioning systems.

The Scope 2 is connected to indirect GHG emissions resulting from the production of electricity, heat, and steam imported and consumed by

[4] European Banking Authority (2020), Risk Dashboard, Data as of Q4 2020, available at https://www.eba.europa.eu/sites/default/documents/files/document_library/Risk%20Analysis%20and%20Data/Risk%20dashboard/Q4%202020/972092/EBA%20Dashboard%20-%20Q4%202020%20-%20footnote%20%281%29.pdf?retry=1.

the organization, as the importer is indirectly responsible for the emissions generated by the supplier for the production of the required energy. The Scope 3 is related to indirect emissions due to the company's activities. This class includes emission sources that are not controlled directly by the company, but whose emissions are indirectly due to company activity. Finally, a Dummy variable that represents the introduction of the Sustainable Finance Action Plan in 2018 is taken in consideration.

1.3.1 *Variables and Univariate Analysis*

Table 1.2 reports the main statistical features of the variables used in our regression analysis.

The definitions of the variables used are provided below.

- *Bank Returns* is computed as the annualized weekly percentage return on banks' shares over the period analyzed. Specifically, it represents the annualized weekly change in price of the sample banks' securities during the period 2014–2020. This ratio indicates the stock's ability to increase or lower the wealth of its shareholders.

Table 1.2 Descriptive statistics

	Obs.	*Mean*	*Std. Dev.*	*Min*	*Max*
Bank returns	315	−0.042	0.341	−0.996	2.559
Size	315	26.348	1.381	23.171	28.742
Book to market	315	1.803	1.787	0	20.164
Profit margin	315	0.149	0.266	−1.430	0.586
Loan to deposit	315	1.065	0.391	0.490	3.005
Debt on equity	315	1.749	2.475	−2.298	10.923
Capital adequacy ratio	315	0.181	0.036	0.104	0.318
Carbon scope 1	315	3.708	0.833	1.380	4.868
Carbon scope 2	315	4.296	0.964	1.259	5.816
Carbon scope 3	315	3.978	0.975	2.040	8.156
Dummy EU	315	0.428	0.495	0	1
Time trend	315	4	2.0031	1	7
Carbon Footprint 1 *x* Dummy EU	315	1.590	1.9132	0	4.8157
Carbon Footprint 2 *x* Dummy EU	315	1.8329	2.2033	0	5.6871
Carbon Footprint 3 *x* Dummy EU	315	1.6837	2.0257	0	5.0977

The table reports summary statistics (mean, medians, standard deviations, min, and max) of the variables used in regressions. The sample period is 2014–2020

Stock performance is usually measured by its fluctuations in price. Higher is the stock price of the bank, better is the performance, meanwhile a decrease in price is often referred to a poor performance (Bolton & Kacperczyk, 2021a).

- *Size* is computed as the natural logarithm of the banks' total assets at the end of the year in Euro. This metric is widely used as firm's factor, since it represents the total volume of business operations or, alternatively, the magnitude of the business activities. This ratio is fundamental in defining bank performance because of the theory of economies of scale, understood as the ability of the company to produce goods or deliver a service at lower cost (Cichello, 2005).
- *Market to Book* is a financial ratio that measures a company's market value relative to its current book value and is felt as an indicator to assess market's perception of a particular value's stock. In addition to that, the ratio reflects if a company's stock is overvalued or undervalued. This variable is one of the most important proxies used by investment advisors and fund managers to measure the value of a company. It is important for investors in their decision-making process, as they need to assess whether the investment is worthwhile (Pioh et al., 2018).
- *Profit Margin* which is one of the most popular profitability ratios to measure how much money the financial institution makes for each dollar of sales generated. The net profit margin is often calculated as the ratio of net income by sales. Both numbers are found on a bank's net income or profit-and-loss statement. Typically expressed as a percentage, profit margin is an important indicator since it represents the capability of a company or business activity to turn sales in profits (Fama & French, 2004).
- *Loan to Deposit* represents the liquidity metric and is expressed as the percentage of bank's total loans to its total deposit. This ratio represents the bank's strength to pay its debits when there is an excessive demand of customer' deposits withdraw. Generally, a lending financial institution that takes deposits should be aware to have a certain amount of liquidity to sustain its normal daily operations. If this ratio is too high, it means that bank lend too much of their deposits, incurring in liquidity shortage mainly in situation of unexpected contingencies. If the ratio is too low, it means that banks lend too few and might have less interest earnings on deposit income (Sari & Septiano, 2020).

- *Debt on Equity* is used to point out the company's financial leverage computed dividing company's total liabilities on shareholders equity. This ratio is very used in finance because it reveals at what percentage the company is financing its growth through its debt. A high Debt on Equity ratio is often referred to high risk, meaning that a company is financing aggressively its growth with debt (Bhandari, 1988).
- *Capital Adequacy Ratio* is a proxy of solvency measured by the ratio of bank's available capital on banks' risk-weighted credit exposure. This is an important indicator because it guarantees the efficiency of the financial system ensuring that banks have enough capital to absorb potential and future losses. A bank with high capital adequacy ratio is considered to have enough capital on reserve to withstand financial losses and unforeseen downturn. In turn, it means that the bank has enough capital available to be above the minimum requirements in terms of reserves (Barnor & Odonkor, 2012).
- *Carbon Scope 1* are the set of all direct emissions that belong to the organization and/are owned by the company. They refer to those greenhouse emissions released into the atmosphere as a consequence of a set of firm-level activities. This category includes all those fuels that produce greenhouses gas emissions that come from combustion (all vehicles owned or controlled by a firm), fugitive installations (leaks of fluorinated greenhouses gases from air conditions systems, heating systems), and industrial and on-site manufacturing process (factory fumes, chemical) (Busch & Lewandowski, 2018).
- *Carbon Scope 2* are all those indirect emissions that derive from the production and generation of purchased energy. Generally speaking, the electricity produced and consumed falls within the Scope 2 borders. In short, emissions are released by the production of the energy and eventually employed by the organization (Lewandowski, 2017).
- *Carbon Scope 3* refers to those indirect emissions that belong to the value chain and companies' business activities, including both upstream and downstream emissions. In short, this kind of scope is strictly linked to the company's operations. The GHG Protocol divides Scope 3 emissions into 15 categories: for this work the most relevant category refers to investments, which are mostly involved in financial institutions' processes. Investments include four main sub-classes, such as equity investments, debt investments, project

finance, managed investments, and client services (Hertwich & Wood, 2018).

- *Dummy EU*, which takes the value of 0 in 2014, 2015, 2016, and 2017, represents the EU Sustainable Finance Action Plan recognized in 2018. Therefore, the dummy takes the value of 1 in 2018, 2019, and 2020, the years in force of the Plan. We include the Sustainable Finance Action Plan as European stimulus to encourage a new sustainable financial strategy shifting away from highly intensive-fossil-fuel-economy and thus improving the allocation of capital to sustainable investments.
- *Time trend* or time index is defined as ordered set of natural numbers. In details 2014 = 1, 2015 = 2, 2016 = 3, 2017 = 4, 2018 = 5, 2019 = 6, 2020 = 7 and it measures the time span between observations. The slope of a time-trend line represents the growing of a variable.

Table 1.3 reports correlations between bank-specific control variables, carbon emissions and the European banks' return between 2014 and 2020.

The model specification utilized in this work is the panel fixed effect analysis, which allows to examine the relationship between banks' carbon emissions and performance. First, the F test statistic was performed (F (44, 252) = 1.50 and Prob > F = 0.01) to let us prefer panel-data regression over the pooled OLS model. The panel data methodology raises the power of empirical analysis, since it combines and mixes information from both cross-section dimensions and time, allowing a greater flexibility in modeling (Greene, 2005). After the F test, we performed the Hausman test as to decide between the fixed or random effects model. The Hausman test showed a statistic chi2(1) = 41.68, P = 0.0012 meaning that the fixed effect panel model best fits our data. At this point, we incorporate both year fixed effects, capturing unobservable factors which are time-variant and may influence banks' stock returns, and country fixed effects, in order to account for unobserved time-invariant heterogeneity across several countries. Robust standard errors are clustered at bank level. Individual time-trend variable is also included, so as to assume whether there is some permanent deterministic pattern across time. In definite, our panel-data regression is as follows:

Table 1.3 Correlation matrix

Correlation matrix	*Bank returns*	*Size*	*Book to market*	*Profit margin*	*Loan to deposit*	*Debt on equity*	*Capital ratio*	*Carbon scope 1*	*Carbon scope 2*	*Carbon scope 3*	*Dummy EU sust*	*Time trend*
Bank returns	1											
Size	0.001	1										
Book to market	−0.321	−0.052	1									
Profit margin	0.423	−0.094	−0.417	1								
Loan to deposit	0.121	−0.220	−0.202	0.335	1							
Debt on equity	0.089	−0.035	−0.181	0.212	0.869	1						
Capital Adequacy Ratio	0.095	0.041	−0.245	0.305	0.270	0.252	1					
Carbon Scope 1	−0.060	0.590	0.102	−0.247	−0.490	−0.387	−0.2686	1				
Carbon Scope 2	−0.008	0.574	0.017	−0.137	−0.365	−0.278	−0.035	0.607	1			
Carbon Scope 3	0.066	0.786	0.012	0.025	−0.222	−0.085	0.005	0.572	0.495	1		
Dummy EU	−0.134	0.023	0.206	0.072	−0.052	−0.148	0.157	0.011	−0.017	0.016	1	
Time Trend	−0.054	0.024	0.212	0.072	−0.069	−0.207	0.256	0.009	−0.025	0.006	0.866	1

$$
\begin{aligned}
RET_{i,t} &= a_0 + a_1 Carbon.Emiss(SCOPE1)_{i,t} \\
&+ a_2 Carbon.Emiss(SCOPE2)_{i,t} + a_1 Carbon.Emiss(SCOPE3)_{i,t} \\
&+ Dummy * Carbon.Emiss(SCOPE1)_{i,t} + Dummy * Carbon.Emiss(SCOPE2)_{i,t} \\
&+ Dummy * Carbon.Emiss(SCOPE3)_{i,t} + a_3 Controls_{i,t-1} \\
&+ \delta t + u_t + e_{i,t}
\end{aligned}
$$

1.4 Empirical Analysis and Results

Table 1.4 provides the results of the main econometric analysis. Consistently with our purposes, our dependent variable is the stock performance of the European banks. By using all three alternative key independent variables in terms of different kind of carbon emission scopes, the empirical analysis supports our hypotheses. We first find consistent evidence for a strong positive impact of our key interested variables on stock returns (Hypothesis 1) and, after the introduction of the Sustainable Action Plan in 2018, we found a negative impact (Hypothesis 2).

What turns out by the analysis conducted shows interesting results. The first explanatory variable statistically important with a negative sign is *Size* (Table 1.4, coefficient = −0.7501, significant at 1%). From a theoretical background, small firm theory conveys that small firms, i.e., those companies which has a small market capitalization, outperform large ones (Gan et al., 2013; Liu, 2006).

Thus, there is evidence that smaller banks have higher expected returns than larger ones, and this is also commonly known under the name of "size premium" (Banz, 1981; Van Dijk, 2011; Zakamulin, 2013).

The Market to Book ratio has a positive and statistically significant coefficient (Table 1.4, coefficient = 0.2273, significant at 1%). This is consistent with that line of research in the economic literature which explains that the market to book ratio has a significant power in describing the cross-sectional changes in stock returns (Asness et al., 2013; Diether et al., 2002; Fama & French, 2015; Hou & Robinson, 2006; Zhang, 2005).

The stock market attributes a higher value to companies which display more earnings power than their assets. It suggests that investors believe the company has excellent future prospects for growth, which in turn increases profits and returns.

Table 1.4 Regression results

	Bank returns
Size	−0.7501***
	(0.1729)
Market to book	0.2273***
	(0.0719)
Profit margin	0.5151***
	(0.927)
Loan to deposit	−0.077
	(0.2002)
Debt on equity	0.0100
	(0.0304)
Capital adequacy ratio	0.6236
	(1.0623)
Carbon scope 1	−0.0335
	(0.0838)
Carbon scope 2	−0.0524
	(0.0444)
Carbon scope 3	0.0961**
	(0.0518)
Dummy EU sustainable plan	1.3673***
	(0.1872)
Time trend	−0.2448***
	(0.0700)
Carbon Footprint 1*Dummy EU	−0.0715
	(0.0577)
Carbon Footprint 2*Dummy EU	0.9670
	(0.0473)
Carbon Footprint 3*Dummy EU	−0.1011***
	(0.0413)
_cons	19.3937
	(4.4349)
Observations	315
Banks	45
R-Squared	0.4332
Country FE	YES
Time FE	YES

The table reports the panel data fixed effect regression results. The sample period goes from 2014 to 2020

Note Standard errors are shown below coefficient estimates

P-values are shown below test statistics

$^{*}p < 0.1$; $^{**}p < 0.05$; $^{***}p < 0.01$

With reference to Profit Margin coefficient (Table 1.4, coefficient = 0.5151, significant at 1%), there is a positive and significant effect on annualized weekly banks' returns.

This relationship should not be surprising since investors prefer companies that offer better profitability ratio, which is translated into higher company prices and, in turn, increases the overall stock returns. This relationship is widely demonstrated in the empirical literature (Allozi & Obeidat, 2016; Endri, 2018; Endri et al., 2019; Er & Vuran, 2012; Fathony et al., 2020; Nurhakim et al., 2016).Our dummy variable is positive and significant, meaning that the introduction of the EU influences the composition of listed banks' returns. The time-trend variable is statistically significant with a negative sign (Table 1.4, coefficient = −0.2448, significant at 1%), meaning that the dependent variable decreases over time, in line with our assumption.

The empirical findings show that, among our key interested independent variables, the Scope 3 emission carbon turns out to be statistically significant. This should not be surprising, since the emissions attributable to the financial and banking industry fall above all in the category of Scope 3: depending on the direction in which they direct their investments, financial operators can in fact increase or reduce their exposure to the fossil fuel sector. The empirical results indicate that, before the introduction of the sustainable guideline introduced by European Commission, banks' Scope 3 carbon emissions have a positive impact on stock returns (Table 1.4, coefficient = 0.0961, significant at 5%), which in turn supports our hypothesis (H1), which could be explained by the fact that banks that include high-carbon-intensive activities have higher carbon exposures (Bolton & Kacperczyk, 2021a, b; Litterman, 2013; Monasterolo & De Angelis, 2020; Weitzman, 2009). This phenomenon is called "carbon premium" (Bolton & Kacperczyk, 2020, 2021a, b; Hsu et al., 2020; Wen et al., 2020). It implies that stock market participants consider carbon emissions as a risk factor when assessing the companies' performance (Matsumura et al., 2014). As a result, investors would require higher compensation for the higher climate risk exposure of these financial institutions (Ilhan et al., 2021).

However, after the introduction of the EU Sustainable Finance Action Plan in 2018, the coefficient of carbon risk factor decreases over time ending up being significantly negative (Table 1.4, coefficient = −0.1011, significant at 1%). The coefficient interaction between the pre and post Sustainable Action Plan in terms of Scope carbon emission 3 is overall negative ($\lambda = 0.0961 - 0.1011 << 0$). In definitive, the negative coefficient of Carbon Footprint 3*Dummy EU (−0.1011***) is really tiny and talking about of a reversal in the effect is really risky but, at the same time, there is a possible feeble indication that the European directive has in a way counterbalance the incentive of banks to pollute.

Indeed, results could indicate that European banks start to internalize the EU guidelines' information, and this could capture the attention of investors to move monetary flows away from polluting banks because of the EU sustainable structural shift.

In a certain sense, the market could rationally discount future cash flows of polluting industries after the European directive, hinting investors to migrate away from carbon-intensive business models. However, what is certain is that the increasing attention to sustainability issues may motivate market participants to allocate their capital to those banks that are environmental-friendly, prompting the decarbonization process, by divesting fossil fuel companies from portfolios (Galema et al., 2008; Henikel et al., 2001; Riedl & Smeets, 2017). This supports in a way our hypothesis (H2).

In sum, the results of this work suggest that there is an indication that pro-environmental EU directives may influence market participants to shift away from firms that include high-carbon-intensive activities (Hong & Kacperczyk, 2009; Pastor & Veronesi, 2012), seeking better financial performance in those that behave ethically in terms of carbon emissions and have low environmental impact (Bauer & Smeets, 2015; Nilsson, 2008). Overall, in the light of new European sustainable guidelines, market participants could select in the long run companies which consider climate change in their business strategy and are better prepared for the transition to a low-carbon economy, seeking for superior returns and low-climate risk exposures (Cheema-Fox et al., 2021; Reghezza et al., 2022).

1.5 Conclusion

The impact of climate change on financial system health is becoming one of the most urgent topics, capturing the attention among scholars, financial institutions and policy makers. Central banks and financial institutions play an important role in combating climate change. Indeed, they help the financial system to be resilient through the transition to a low-carbon economy by providing transparent information to all market participants and, at the same time, by reducing greenhouses emissions.

This study tried to explain the nature of the relationship between the 45 main European carbon banks' emissions and relative stock returns over the period that goes from 2014 to 2020. These banks operate in 17 European countries. To the best of our knowledge, little is known about how carbon emissions affect European banking sector in terms of stock returns, and therefore, our study is aimed at solving this question. The originality of this study is twofold. First, what comes out is that before the introduction of European sustainable regulations such as the Sustainable Finance Action Plan, investors would claim a "carbon premium," i.e., greater stock price returns for those European financial institutions that are carbon inefficient. This is mainly due to the perception of the investors for the high climate risk exposure. Rauh et al. (2011) call this an "insurance-like protection against negative events." Secondly, the tiny negative relationship that exist between European banks performance and carbon emissions after the introduction of the European directive in 2018 ($\lambda = 0.0961 - 0.1011 << 0$) reveals a possible indication for stock market investors to address their investments toward sustainable portfolios, aligned with EU guidelines.

Indeed, nonetheless the negative coefficient, talking about a definitive reversal in the effect, is difficult given the empirical results, but at the same time these outcomes could be seen as a possible suggestion and indication for investors to address their portfolios toward low-carbon emission assets in the long run, in line with the increasing awareness of the regulatory European bodies toward sustainability. This is widely demonstrated by the growing mass of investors who are acting against climate change, creating sustainable portfolios for a low-carbon future and prompting the so-called portfolio decarbonization process.

Then, the findings of the work could add further information on how the financial market may perceives banks' carbon activities in the next years, raising the understanding of investors and market participants on allocating capital toward corporate environmental investments. What is certain is that investors will be exposed to transition-carbon risk in the next decades. Given the nature of the stock markets to look forward, it is natural to ask to what extent the carbon-climate risk is incorporated in stock returns.

This research is not without some limitations. We focus on European area banks only, not taking in account other financial institutions headquartered in other countries such as US, South America, or Asia. It might be interesting to extend the analysis to international banks as well. Furthermore, we do not consider the single business models of the banks taken in analysis and strategies of individual banks toward sustainability. In this view, we presume these may be some food for thought for future research.

References

Aggarwal, R., & Dow, S. (2013). Corporate governance and business strategies for climate change and environmental mitigation. In *Entrepreneurship, finance, governance and ethics* (pp. 315–340). Springer.

Allozi, N. M., & Obeidat, G. S. (2016). The relationship between the stock return and financial indicators (profitability, leverage): An empirical study on manufacturing companies listed in Amman Stock Exchange. *Journal of Social Sciences (Coes&Rj-Jss), 5*(3), 408–424.

Alvarez, I. G. (2012). Impact of CO2 emission variation on firm performance. *Business Strategy and the Environment, 21*(7), 435–454.

Asness, C. S., Moskowitz, T. J., & Pedersen, L. H. (2013). Value and momentum everywhere. *The Journal of Finance, 68*(3), 929–985.

Bank of England. (2015). *The impact of climate change on the UK insurance sector*. A Climate Change Adaptation Report by the Prudential Regulation Authority. https://www.bankofengland.co.uk/-/media/boe/files/prudential-regulation/publication/impact-of-climate-change-on-the-uk-insurance-sector.pdf

Bank of England. (2021). https://www.bankofengland.co.uk/climate-change

Banz, R. W. (1981). The relationship between return and market value of common stocks. *Journal of Financial Economics, 9*(1), 3–18.

Barnor, C., & Odonkor, T. A. (2012). Capital adequacy and the performance of Ghanaian banks. *Journal of Business Research, 6*(1–2), 105–117.

Bauer, R., & Smeets, P. (2015). Social identification and investment decisions. *Journal of Economic Behavior & Organization, 117*, 121–134.

Bernardini, E., Di Giampaolo, J., Faiella, I., & Poli, R. (2021). The impact of carbon risk on stock returns: Evidence from the European electric utilities. *Journal of Sustainable Finance & Investment, 11*(1), 1–26.

Bhandari, L. C. (1988). Debt/equity ratio and expected common stock returns: Empirical evidence. *The Journal of Finance, 43*(2), 507–528.

Boiral, O., Henri, J. F., & Talbot, D. (2012). Modeling the impacts of corporate commitment on climate change. *Business Strategy and the Environment, 21*(8), 495–516.

Bolton, P., & Kacperczyk, M. (2020). *Carbon premium around the world* (CEPR Discussion Papers No. 14567).

Bolton, P., & Kacperczyk, M. (2021a). Do investors care about carbon risk? *Journal of Financial Economics, 142*(2), 517–549.

Bolton, P., & Kacperczyk, M. T. (2021b). *Global pricing of carbon-transition risk* (NBER Working Paper w28510).

Busch, T., & Hoffmann, V. H. (2011). How hot is your bottom line? Linking carbon and financial performance. *Business & Society, 50*(2), 233–265.

Busch, T., & Lewandowski, S. (2018). Corporate carbon and financial performance: A meta-analysis. *Journal of Industrial Ecology, 22*(4), 745–759.

Chapple, L., Clarkson, P. M., & Gold, D. L. (2013). The cost of carbon: Capital market effects of the proposed emission trading scheme (ETS). *Abacus, 49*(1), 1–33.

Cheema-Fox, A., LaPerla, B. R., Serafeim, G., Turkington, D., & Wang, H. S. (2021). Decarbonization factors. *The Journal of Impact and ESG Investing*.

Cichello, M. S. (2005). The impact of firm size on pay–performance sensitivities. *Journal of Corporate Finance, 11*(4), 609–627.

Delmas, M. A., Nairn-Birch, N., & Lim, J. (2015). Dynamics of environmental and financial performance: The case of greenhouse gas emissions. *Organization & Environment, 28*(4), 374–393.

Diether, K. B., Malloy, C. J., & Scherbina, A. (2002). Differences of opinion and the cross section of stock returns. *The Journal of Finance, 57*(5), 2113–2141.

Endri, E. (2018). Factors determine stock return of livestock feed companies: Common effect model analysis. *International Journal of New Technology and Research, 4*(5), 263051.

Endri, E., Dermawan, D., Abidin Z., & Riyanto, S. (2019). Effect of financial performance on stock return: Evidence from the food and beverages sector. *International Journal of Innovation, Creativity and Change, 9*(10), 335–350.

Environmental Protection Agency. (2021). Climate change indicators: Greenhouse gases. https://www.epa.gov/climate-indicators/greenhouse-gases

Er, Ş, & Vuran, B. (2012). Factors affecting stock returns of firms quoted in ISE market: A dynamic panel data approach. *International Journal of Business and Social Research (IJBSR), 2*(1), 109–122.

European Banking Authority. (2021). *Mapping climate risk: Main findings from the EU-wide pilot exercise* (EBA/Rep/2021/11, 21 May 2021, Report).

https://www.eba.europa.eu/sites/default/documents/files/document_library/Publications/Reports/2021/1001589/Mapping%20Climate%20Risk%20-%20Main%20findings%20from%20the%20EU-wide%20pilot%20exercise%20on%20climate%20risk.pdf

European Central Bank. (2021). https://www.ecb.europa.eu/ecb/climate/html/index.en.html

European Environmental Agency. (2012). *Climate change, impacts and vulnerability in Europe* (EEA Reports, No 12/2012).

European Environmental Agency. (2021). *Trends and projections in Europe 2021* (European Environment Agency reports).

European Systemic Risk Board. (2016, February). *Too late, too sudden: Transition to a low-carbon economy and systemic risk* (Reports of the Advisory Scientific Committee, No. 6). https://www.esrb.europa.eu/pub/pdf/asc/Reports_ASC_6_1602.pdf

Faiella, I., & Malvolti, D. (2020). *The climate risk for finance in Italy* (Bank of Italy Occasional Paper (545)).

Faiella, I., & Mistretta, A. (2015). Spesa energetica e competitività delle imprese italiane. *Spesa energetica e competitivitá delle imprese italiane*, 85–121.

Fama, E. F., & French, K. R. (1995). Size and book-to-market factors in earnings and returns. *The Journal of Finance, 50*(1), 131–155.

Fama, E. F., & French, K. R. (2004). New lists: Fundamentals and survival rates. *Journal of Financial Economics, 73*(2), 229–269.

Fama, E. F., & French, K. R. (2015). A five-factor asset pricing model. *Journal of Financial Economics, 116*(1), 1–22.

Fathony, M., Khaq, A., & Endri, E. (2020). The effect of corporate social responsibility and financial performance on stock returns. *International Journal of Innovation, Creativity and Change, 13*(1), 240–252.

Federal Reserve. (2021). https://www.federalreserve.gov/newsevents/speech/brainard20211007a.htm

Ferrell, A., Liang, H., & Renneboog, L. (2016). Socially responsible firms. *Journal of Financial Economics, 122*(3), 585–606.

Galema, R., Plantinga, A., & Scholtens, B. (2008). The stocks at stake: Return and risk in socially responsible investment. *Journal of Banking & Finance, 32*(12), 2646–2654.

Gallego-Álvarez, I., García-Sánchez, I. M., & da Silva Vieira, C. (2014). Climate change and financial performance in times of crisis. *Business Strategy and the Environment, 23*(6), 361–374.

Gallego-Álvarez, I., Segura, L., & Martínez-Ferrero, J. (2015). Carbon emission reduction: The impact on the financial and operational performance of international companies. *Journal of Cleaner Production, 103*, 149–159.

Gan, C., Hu, B., & Li, Z. (2013). An empirical cross-section analysis of stock returns on the Chinese A-share stock market. *Investment Management and Financial Innovations, 10*(1), 127–136.

Greene, W. (2005). 4 Models for panel data. *Foundations and Trends in Econometrics, 1*(2), 152–161.

Hatakeda, T., Kokubu, K., Kajiwara, T., & Nishitani, K. (2012). Factors influencing corporate environmental protection activities for greenhouse gas emission reductions: The relationship between environmental and financial performance. *Environmental and Resource Economics, 53*(4), 455–481.

Heinkel, R., Kraus, A., & Zechner, J. (2001). The effect of green investment on corporate behavior. *Journal of Financial and Quantitative Analysis, 36*(4), 431–449.

Hertwich, E. G., & Wood, R. (2018). The growing importance of scope 3 greenhouse gas emissions from industry. *Environmental Research Letters, 13*(10), 104013.

Hong, H., & Kacperczyk, M. (2009). The price of sin: The effects of social norms on markets. *Journal of Financial Economics, 93*(1), 15–36.

Hong, H., Li, F. W., & Xu, J. (2019). Climate risks and market efficiency. *Journal of Econometrics, 208*(1), 265–281.

Hou, K., & Robinson, D. T. (2006). Industry concentration and average stock returns. *The Journal of Finance, 61*(4), 1927–1956.

Hsu, P. H., Li, K., & Tsou, C. Y. (2020). *The pollution premium* (Working paper). Available at SSRN 3508808.

Ilhan, E., Sautner, Z., & Vilkov, G. (2021). Carbon tail risk. *The Review of Financial Studies, 34*(3), 1540–1571.

IPCC. (2014). *Climate change 2014: Mitigation of climate change. Contribution of working group III to the fifth assessment report of the intergovernmental panel on climate change* (O. Edenhofer, R. Pichs-Madruga, Y. Sokona, E. Farahani, S. Kadner, K. Seyboth, A. Adler, I. Baum, S. Brunner, P. Eickemeier, B. Kriemann, J. Savolainen, S. Schlömer, C. von Stechow, T. Zwickel, & J. C. Minx, Eds.). Cambridge University Press. https://www.ipcc.ch/site/assets/uploads/2018/02/ipcc_wg3_ar5_frontmatter.pdf

IPCC. (2018). *Global warming of 1.5°C. An IPCC special report on the impacts of global warming of 1.5°C above pre-industrial levels and related global greenhouse gas emission pathways, in the context of strengthening the global response to the threat of climate change, sustainable development, and efforts to eradicate poverty* [V. Masson-Delmotte, P. Zhai, H.-O. Pörtner, D. Roberts, J. Skea, P. R. Shukla, A. Pirani, W. Moufouma-Okia, C. Péan, R. Pidcock, S. Connors, J. B. R. Matthews, Y. Chen, X. Zhou, M. I. Gomis, E. Lonnoy, T. Maycock, M. Tignor, & T. Waterfield, Eds.). https://www.ipcc.ch/site/assets/uploads/sites/2/2019/06/SR15_Full_Report_High_Res.pdf

IRENA. (2017). *Stranded assets and renewables: How the energy transition affects the value of energy reserves, buildings and capital stock.* International Renewable Energy Agency (IRENA). www.irena.org/remap

Karydas, C., & Xepapadeas, A. (2019). *Pricing climate change risks: CAPM with rare disasters and stochastic probabilities* (CER-ETH Working Paper Series Working Paper, 19, 311).

Krueger, P., Sautner, Z., & Starks, L. T. (2020). The importance of climate risks for institutional investors. *The Review of Financial Studies, 33*(3), 1067–1111.

Lewandowski, S. (2017). Corporate carbon and financial performance: The role of emission reductions. *Business Strategy and the Environment, 26*(8), 1196–1211.

Litterman, B. (2013). What is the right price for carbon emissions. *Regulation, 36*, 38.

Liu, W. (2006). A liquidity-augmented capital asset pricing model. *Journal of Financial Economics, 82*(3), 631–671.

Matsumura, E. M., Prakash, R., & Vera-Munoz, S. C. (2014). Firm-value effects of carbon emissions and carbon disclosures. *The Accounting Review, 89*(2), 695–724.

Monasterolo, I., & De Angelis, L. (2020). Blind to carbon risk? An analysis of stock market reaction to the Paris Agreement. *Ecological Economics, 170*, 106571.

Nilsson, J. (2008). Investment with a conscience: Examining the impact of pro-social attitudes and perceived financial performance on socially responsible investment behaviour. *Journal of Business Ethics, 83*(2), 307–325.

Nurhakim, A., Yunita, I., & Iradianty, A. (2016). The effect of profitability and inflation on stock return at pharmaceutical industries at Bei in the period of 2011–2014. In *First International Conference on Advanced Business and Social Sciences* (Vol. 2, No. 2, pp. 202–210).

OECD. (2018). *A chain reaction: Disruptive innovation in the electricity sector*. www.oecd.org/competition/A-chain-reaction-disruptive-innovation-in-the-electricity-sector.htm

Oestreich, A. M., & Tsiakas, I. (2015). Carbon emissions and stock returns: Evidence from the EU emissions trading scheme. *Journal of Banking & Finance, 58*, 294–308.

Pastor, L., & Veronesi, P. (2012). Uncertainty about government policy and stock prices. *The Journal of Finance, 67*(4), 1219–1264.

Pindyck, R. S. (2013). Pricing carbon when we don't know the right price. *Regulation, 36*(2), 43–46.

Pioh, H. T., Tommy, P., & Sepang, J. L. (2018). The effect of debt to equity ratio, earning per share, and return on assets on firm value in the food and beverages sub-sector in the Indonesia stock exchange. *Journal of EMBA Accounting, 6*(4), 3018–3027.

Rauh, V., Arunajadai, S., Horton, M., Perera, F., Hoepner, L., Barr, D. B., & Whyatt, R. (2011). Seven-year neurodevelopmental scores and prenatal exposure to chlorpyrifos, a common agricultural pesticide. *Environmental Health Perspectives, 119*(8), 1196–1201.

Reghezza, A., Altunbas, Y., Marques-Ibanez, D., d'Acri, C. R., & Spaggiari, M. (2022). Do banks fuel climate change? *Journal of Financial Stability, 62*, 101049.

Riedl, A., & Smeets, P. (2017). Why do investors hold socially responsible mutual funds? *The Journal of Finance, 72*(6), 2505–2550.

Sari, L., & Septiano, R. (2020). Effects of intervening loan to deposit ratio on profitability. *Journal of Accounting and Finance Management, 1*(5), 228–241.

Siddique, M. A., Akhtaruzzaman, M., Rashid, A., & Hammami, H. (2021). Carbon disclosure, carbon performance and financial performance: International evidence. *International Review of Financial Analysis, 75*, 101734.

Testa, F., Iraldo, F., & Frey, M. (2011). The effect of environmental regulation on firms' competitive performance: The case of the building & construction sector in some EU regions. *Journal of Environmental Management, 92*(9), 2136–2144.

Tol, R. S. (2008). The social cost of carbon: trends, outliers and catastrophes. *Economics, 2*(1).

Trinks, A., Mulder, M., & Scholtens, B. (2020). An efficiency perspective on carbon emissions and financial performance. *Ecological Economics, 175*, 106632.

U.S. Global Change Research Program. (2021). https://www.globalchange.gov/browse/reports/our-changing-planet-fy-2021

Van Dijk, M. A. (2011). Is size dead? A review of the size effect in equity returns. *Journal of Banking & Finance, 35*(12), 3263–3274.

Weitzman, M. L. (2009). On modelling and interpreting the economics of catastrophic climate change. *The Review of Economics and Statistics, 91*(1), 1–19.

Wen, F., Wu, N., & Gong, X. (2020). China's carbon emissions trading and stock returns. *Energy Economics, 86*, 104627.

Zakamulin, V. (2013). Forecasting the size premium over different time horizons. *Journal of Banking & Finance, 37*(3), 1061–1072.

Zhang, L. (2005). The value premium. *The Journal of Finance, 60*(1), 67–103.

Ziegler, A., Schröder, M., & Rennings, K. (2007). The effect of environmental and social performance on the stock performance of European corporations. *Environmental and Resource Economics, 37*(4), 661–680.

CHAPTER 2

Dissecting the European ESG Premium vs the US: Is It All About Non-financial Reporting?

Rocco Ciciretti, Ambrogio Dalò, and Giovanni Ferri

2.1 Introduction

The European Union has been moving toward sustainable transition for some time and the pace has now accelerated, first with the European Green Deal (EGD) and then with strong green (e.g., Next Generation

We are particularly grateful to Flavia Nuccitelli for her very valuable work as a research assistant on this project.

G. Ferri (✉)
Department of Law, Economics, Politics and Modern Languages, LUMSA University of Rome, Rome, Italy
e-mail: g.ferri@lumsa.it

R. Ciciretti
Department of Economics and Finance and RCEA, Tor Vergata University, Rome, Rimini, Italy
e-mail: rocco.ciciretti@uniroma2.it

P. Wachtel et al. (eds.), *Creating Value and Improving Financial Performance*, Palgrave Macmillan Studies in Banking and Financial Institutions, https://doi.org/10.1007/978-3-031-24876-4_2

EU) and social (e.g., SURE) sustainable connotations included in the relaunch programs from the COVID-19 crisis. Optimists think that, in addition to being necessary to safeguard the future, becoming sustainable gives Europe a real competitive advantage in a world where demand will increasingly shift in consumption toward green products and in investment toward financial assets issued by sustainable businesses. Pessimists, on the other hand, believe that EU policies for sustainable transition impose huge and unjustified costs. Who is right? Let's try to answer by evaluating whether European companies have gained an advantage over US companies, the latter having been also held back by Trump's parenthesis, in a very emblematic and dynamic segment, that of sustainable finance.

Sustainable finance consists of various components: the most dynamic is that of SRI funds—Sustainable & Responsible Investment Funds—whose investment strategies typically use ESG ratings. This is an area of strong growth and the estimates in circulation place ESG investments at stratospheric levels: They could reach $35 trillion in 2025. Therefore, given this impetuous growth, today competitiveness also depends on how companies are positioned to tap into sustainable finance markets. In fact, for a company, the ESG rating measures performance in managing environmental risks (E—Environmental), social ones (S—Social), and those of its own administration (G—Governance).

Well, comparing US companies listed in the S&P500 with a similar group of listed European companies, we show that on average EU-listed companies enjoy 14% higher ESG ratings than US companies (64.43 for European companies versus 56.37 for the US ones). If we want to break it down, the advantage of European companies is maximum in the E—Environmental component (+22.5%; 63.08 against 51.48), intermediate in the S—Social component (+16.0%; 68.35 against 58.90), and negligible in the G—Governance component (+0.4%; 59.88 against 59.66). Furthermore, it is shown that the EU advantage does not depend on the different sectoral composition of EU companies which, at the most, would assign European companies +0.64% compared to US companies.

A. Dalò
Faculty of Economics and Business, University of Groningen, Groningen, The Netherlands
e-mail: a.dalo@rug.nl

Instead, the EU advantage is found to be related to the fact that EU companies more often than their US counterparts are committed to providing sustainability reports and provide better quality sustainability reports. In fact, even before the enforcement achieved in 2018 of Directive no. 2014/95/EU on non-financial reporting, between 2011 and 2017 almost two thirds of the European listed companies considered in the analysis (62.29%) prepared sustainability reports according to the GRI (Global Reporting Initiative) scheme while less than half (44.86%) of the US ones did that. Furthermore, by evaluating the quality of the GRI reports—that is, how much companies choose more or less advanced reporting methods—it is estimated that the average European report is 8.51% more accurate than the US one.

These results confirm that EU policies in favor of sustainability can bear fruit by offering European companies advantages in accessing sustainable finance. So, at least in this area, the optimists seem to be right.

The rest of the chapter is structured as follows. Section 2.2 recaps the perception of this EU vs US gap as we can find in the extant literature. Section 2.3 develops the two main hypotheses to be tested. Then, Sect. 2.4 outlines the empirical study undertaken with a methodology section—explaining how the data was collected and analyzed—and the presentation of our core findings based on descriptive evidence but also on econometric analysis. Finally, Sect. 2.5 summarizes our main results, warns about some caveats, and articulates possible new avenues for future research.

2.2 The Perception of the EU–US Gap in Terms of ESG Ratings by the Extant Literature

In recent years, we can observe that, on average, ESG scores are higher for European vs US companies. This fact is mentioned in various works, reports, and working papers that are quite heterogeneous in terms of ratings used, nature of the observer—whether scholars or investment practitioners—and period considered. In a study center on developed markets, Peiró-Signes and Segarra-Oña (2013) examine ASSET4's ratings for the period 2006–2010 and find that European companies—being the best performing regional group of all—have average and median ESG scores significantly higher than US ones. Besides, for the total ESG scores, the gap attains for both the environmental and social scores—while the governance score is slightly higher for US companies. [The article reports these average ratings: European companies vs US companies: Total ESG score 0.685 vs 0.526; E score 0.632 vs 0.409; S score 0.687 vs 0.443; G score 0.633 vs 0.724.] Following this finding, Bannier et al. (2019) focus expressly on a US-Europe comparison, report further evidence, from data by the same provider on a longer period (2003–2017), and narrow on the highest-capitalized listed companies. They find that the mean ESG total score is lower for the US (50.8) vs the European sample (56.7), with better mean environmental and social scores for Europe and analogous average governance score. In Nitsche and Schröder's (2019) research on SRI funds by three different ratings (Oekom, Sustainalytics, and ASSET4), the fact that EU companies generally receive higher scores is used as an assumption to explain skewed results in the reported mean scores. Using company-level scores from Sustainalytics, Morningstar's (2019) *Sustainability Atlas*, shows that—while the US lags behind in the fourth quintile—Europe is home to the most sustainable countries and adds: "This is somewhat expected, since those nations have always been ahead of the curve on this front."

For practitioners, asset management company KBI Global Investors (2019) points at this trend in its broader regional analysis of ESG performance, by MSCI ESG ratings for 2018. The study shows that Europe has a higher concentration of stocks with a higher ESG score than North America. In particular, in Europe (North America), 71.4% (35%) of companies achieved above-average ESG scores. On Sustainalytics and MSCI data in the context of an analysis on sustainable bonds, Barclays (2018) notes that European issuers tend to have higher ESG ratings than US issuers. Also Hill (2020) supports Barclays's findings to draw conclusions in its review on sustainable investment. In addition, the asset manager Nordea Markets (2018), in a paper on the potential for Scandinavian companies from the increased trends in SRI, shows the average ESG scores for the regional groups of Nordic, European, and North American companies (MSCI data over 2005–2017). Besides illustrating other data, the report clearly shows that the average score for European companies is stably and significantly higher than the American one over the whole period. Some analysts stress the same issue in a 'white paper' for the asset management company Legg Mason, showing how European stocks take over American ones in the highest deciles of the score distribution according to all three ratings used—MSCI, Thomson Reuters/ASSET4, Sustainalytics—over 2012–2018 (LaBella et al., 2019).

In each of these cases, the issue arises in comparing ESG ratings across regions. As said above, a geographic bias has been acknowledged by critics, some of whom identify an advantage for European companies due to EU rules on non-financial disclosure. More in detail, as explained by Barclays's report, as the lack of disclosure can depress ESG ratings, companies based in Europe can more easily obtain high ESG ratings since they have to follow stricter non-financial disclosure rules. Also, LaBella et al. (2019) explain their results by saying: "rating agencies show a clear bias favoring developed markets outside of the US, particularly European companies [...]. The source of this bias may not fully reflect the quality of ESG practices, but rather the existence and quality of formal reporting requirements in various jurisdictions." Along the same lines, Beloe (2016) claims: "European companies [...] tend to have a culture of greater disclosure on ESG issues and in some countries ESG disclosure is mandatory. As a consequence, ESG ratings tend to be much higher in Europe. In one case, the average percentile score for European companies is nearly 20 percentage points higher than the average percentile score in the US. Perhaps European companies are actually better at managing ESG issues, but this gap is so huge that at least some of it is likely to be due to differences in disclosure."

Doyle (2018) reports a specific example which helps visualize how the geographic bias can affect ESG comparisons in US vs Europe. By focusing on auto manufacturers, he compares Sustainalytics scores for the BMW Group (Europe) and Tesla (US) in 2016. BMW has a high rating (74, ranked at 93rd percentile) despite many controversies on anti-competitive and illegal marketing practices, violations of intellectual property, and of employee and human rights, even of animal rights, and allegations of collusion with other European carmakers on various technologies and systems to evade environmental and safety regulations. Instead, Tesla, despite being the world leader in the technology to curb cars' carbon emissions, has a low rating (54, ranked at 38th percentile) below every European carmaker, even below those ones facing major environmental violations. Though this is anecdotal evidence, the example is a warning light on the true lead of European companies in ESG practices.

Although pundits seem to agree that the gap in mean ESG scores in EU vs US reflects a geographical bias, partly due to different regulations, we still need to fully understand the linkages. For instance, authors neither provide evidence proving this causality, nor cite other relevant supportive works. Also, the cited studies hinge on empirical analyses averaging ESG ratings of many companies, whose reporting practices are not put in context to allow a direct connection. For these reasons, this intuition cannot be taken for granted and we will investigate it in the hypotheses through this work.

In conclusion, the perceived gap in ESG scores questions whether European companies are actually 'more sustainable' than their US counterparts. Indeed, in practice and in academia, there is growing reliance on ESG ratings to reflect the level of 'sustainability' of a company, and by this reasoning, this gap in the average ESG scores should automatically imply that European companies are more sustainable, i.e., that their ESG performance is systematically better than the US ones. At the same time, we have reviewed some issues about origins, methodological features, and challenges of ESG ratings which allow one to assume that the inherent sustainability of companies, or ESG performance, may not be the only factor explaining this difference with respect to two regions which are comparable in terms of market development and of CSR traditions.

Therefore, the next sections will address the issue of the gap between average ESG scores of European vs US companies. In particular, two hypotheses are presented to explain the possible origin of this gap, and then an empirical section will bring some findings building on these hypotheses.

2.3 The ESG Score Gap Between EU and US Companies: Hypothesis Development

Here we focus on two among the possible factors behind the identified 'gap' in mean ESG scores of EU vs US companies: (i) differences in the sectoral mix; (ii) differences in sustainability disclosure. These two hypotheses build on the literature and on recent developments of the ESG rating industry. While logically separate, (i) does not exclude (ii) they may reinforce or abate each other.

2.3.1 *Sectoral Composition of Regional Groups of Companies*

A first possible cause of the gap between the mean ESG scores of the two areas concerns the industries or sectors to which the rated companies belong. Indeed, for any given rating agency, the overall universe of all rated companies is distributed across several industries. This distribution can be uneven in terms of sectoral composition, because—as mentioned above—the issuance of ESG ratings is determined by investors' demand and does not imply particular efforts of representativeness.

2.3.1.1 Relevance of the Industry/Sector

The importance of duly accounting for different industries/sectors in interpreting ESG scores emerges for both scholars and practitioners. Capelle-Blancard and Petit (2015, p. 4) identify the issue studying how ESG ratings operationalize CSR concepts. In their words, "Previous work [in the literature] has implied that 'one size fits all'. This hypothesis seems debatable at least: Environmental issues, for example, are likely more important in the Oil & Gas than in the Banking sector. [...] Unsurprisingly, banks are mainly criticized for their bad corporate governance, while they have good environmental reputations. Conversely, firms in the Basic-resource and Oil & Gas sectors are mostly criticized for environmental damage. Last, large retailers (included in the Consumer goods and services sector) have a poor social record. Composite equal-weight scores thus misrepresent the differences between sectors." Indeed, they claim that the key question is how to apply the weights to the industries.

Using industry weights is an evolution on early contributions in the CSR literature which insisted on the uniqueness of the industries, in terms of internal competencies or external pressure, as an obstacle to meaningful comparisons across studies or to generalizations beyond the boundaries of a specific study (Griffin & Mahon, 1997; Rowley & Berman, 2000). At the same time, weights can offer nuances and overcome the rigid separation into 'controversial' sectors vs 'non-controversial' ones, the former including 'sinful' industries such as tobacco, gambling, alcohol, and industries involved with emerging environmental, social, or ethical issues, i.e., weapons, nuclear, oil, cement, and biotech (Cai et al., 2012). Indeed, by using weights ESG ratings manage to capture the CSR efforts made by companies in these areas, previously discarded a priori, while reflecting their possibly problematic nature (e.g., Ktat, 2017). Hence, the advantage of weighting according to the industries is that it still allows large cross-sectional comparisons and, simultaneously, allows to maintain the particularities of different sectors to the eyes of the researchers (Capelle-Blancard & Petit, 2015).

Practitioners seem interested too in integrating industries in composite scores. Already in 2010, the first surveys on investment professionals by the consulting firm SustainAbility (in the research project *Rate the Raters*) noted that "evaluating companies across sector, geography, revenue and different issues is very difficult [...]; many ratings insufficiently consider the context of certain companies, industries and issues" (SustainAbility, 2010a). In their inventory of rating providers, SustainAbility (2010b) claims that "it is difficult—perhaps meaningless—to compare companies from different sectors and geographies on the same set of criteria. Ranking an investment bank against a food retailer against a pharmaceutical company across a common set of criteria is a considerable challenge, as each type of company faces a different set of key issues." But, they noted, "the 'universal' rating—one which spans multiple issues, industries and/or regions—remains the norm"; their best practices advise that ESG raters "[offer] details on criteria, weightings and scoring schemes" and that "ratings must be based predominantly on sector-specific criteria and weightings" (SustainAbility, 2011).

The relevance of such recommendations has become evident over time, with practice evolving along those lines. With the development of the ESG rating industry, today the most prominent agencies do attribute a 'universal,' overall rating to each company, but most raters also provide separated scores by the three themes (E, S, G) and often add further details in the form of reports (e.g., company, country, and industry reports), highlighting pertinent ESG issues and other qualitative information which can help to put the rating in context. While the way to report this deeper research can vary, the ultimate focus is on synthetic scores to convey the main findings and allow comparing rated entities.

In particular, from the analyses *by industry* carried out by rating agencies, we know the mean ESG score of any industry: It is obtained by aggregating the ESG scores of all the companies which fall in the industry and belong to the universe of the same rater. The raters themselves often calculate industry averages, with the specific aim to capture aggregate performances and identify time trends.

Table 2.1 shows the average ESG scores of a set of industries based on the sector reports and articles made publicly available by three different rating agencies: Vigeo Eiris, Sustainalytics, and ASSET4. [Note that scales vary by ESG rater: Each different rating can be mapped between 0 and 100 but each agency has special categories to rank the numeric value, hence a comparison across scores of different providers is not meaningful; the focus here is on the ranking of different industries according to their average ESG score, conveyed by the order of each column.] We chose 8 industries—which do not exhaust the areas investigated by the three raters—based on the availability of public reports and on the correspondence of industry categories. [In particular, they are the totality of industries covered by Sustainalytics' materials on the rater's website. By searching the same industry coverage for the other two raters, a common basis has been found around these 8 groups.] The ranking of these mean ESG scores across raters identifies some industry patterns. For instance, the ranks of Utilities and Household/Personal Products vs Real Estate are, respectively, high and low.

The similarities in industry rankings we can see in Table 2.1 may reflect similar ways to interpret ESG-related risks, as integrated in an agency's rating methodology through its weighting. The ESG Risk Atlas by S&P Global offers a comprehensive view on this, reviewing the relative environmental and social exposures of a wide range of industries (Fig. 2.1).

Although other agencies might provide different views, the example of systematization above points out that, irrespective of the rating agency, single industries face inherently different ESG issues impacting on their weighting system, ultimately playing a role in the scores of individual companies. Given the relevance of industries for ESG assessments, we must complain that the studies, cited above, identifying the Europe–US

Table 2.1 Industries/sectors ranked by average ESG score, sorted by rating agency

Vigeo Eiris	*Sustainalytics*	*ASSET4*
Utilities (35.5)	Household/Personal Products (65)	Aerospace and Defense (73.8)
Household/Personal Products (35)	Aerospace and Defense (63)	Utilities (59.4)
Mining and Metals (33.6)	Utilities (61)	Household/Personal Products (55.8)
Banks (32)	Mining and Metals (59)	Pharmaceuticals and Biotechnology (51.5)
Aerospace and Defense (31.2)	Banks (58)	Insurance (50)
Insurance (29.7)	Insurance (57.8)	Banks (46.4)
Real Estate (29.1)	Real Estate (57)	Mining and Metals (44.7)
Pharmaceuticals and Biotechnology (27)	Pharmaceuticals and Biotechnology (56.7)	Real Estate (33.5)

Source Our reworking of data from the agencies' sector reports

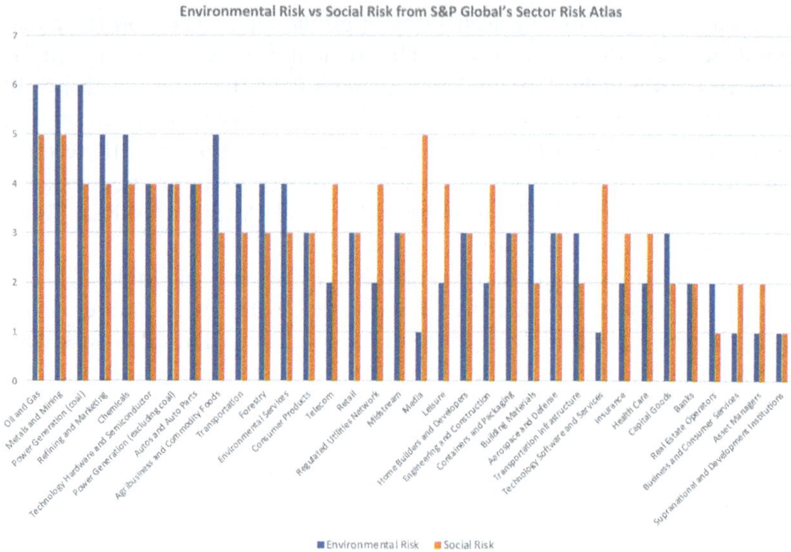

Fig. 2.1 Environmental risk vs social risk from S&P Global's sector risk Atlas (*Source* Adapted by the authors based on S&P Global [2019])

gap in average ESG scores generally use the overall-ESG scores to make the comparison—in some cases observing also the individual themes E, S, and G—failing to consider how possibly different industry mixes in the two regions might affect the comparison.

2.3.1.2 Hypothesis Development: Sectoral Composition

Given that industries/sectors play a role in ESG ratings, the gap between European and US companies' scores might somehow depend on a different sectoral composition of the two regional groups of companies. In accordance with the findings emerging from literature and practice, we formulate the following hypothesis:

Hypothesis 1 A homogeneous distribution across industries/sectors might result in a smaller gap in average ESG ratings between European and US companies.

2.3.2 *Sustainability Reporting Practices*

A second factor which may contribute to the gap between the average ESG scores of European and US companies corresponds to their practices of sustainability reporting (also called 'sustainability disclosure,' 'ESG reporting,' or 'ESG disclosure'). We investigate this element to deepen the intuition proposed by some authors who have observed the geographical gap and have claimed that different regulations and traditions on non-financial disclosure lead to better sustainability reporting by European companies, ultimately delivering higher ESG ratings than at their US counterparts. Because this link has only been implied so far, here we explore and deepen its arguments and ultimately formulate a hypothesis for the empirical analysis.

2.3.2.1 *The Regulatory Framework: Mandatory vs Voluntary Sustainability Reporting*

As concerns the regulatory framework for corporate sustainability, sustainability reporting is mandatory in most European countries, while in the US it is still largely voluntary. In European legislation, Directive n. 2014/95/EU, the subsequent norms and the following guidelines have recently promoted a marked improvement in sustainability reporting especially for larger companies or groups. By amending to the Accounting Directive 2013/34/EU, the 2014 Directive has made it mandatory to report non-financial information for companies whose size is above a certain threshold (i.e., more than 500 employees, and either total assets exceeding EUR (euro—€) 20 million or a net turnover exceeding EUR 40 million). This requires a suitable report focusing on the business model, policies pursued or implemented, sustainability risks, and Key Performance Indicators (KPIs), with a minimum coverage of: environmental and social matters, respect for human rights, anti-corruption and bribery issues, and diversity on the boards of directors. The underlying rationale is to give an incentive to, notably prominent, companies behave more responsibly, and at the same time, increasing attention is addressed to financial investors and other external stakeholders. All this context facilitates generating and elaborating ESG data (Taliento et al., 2019). As to materiality, the threshold for ESG reported data is not necessarily linked to financial considerations: according to Article 1 of the Directive, in fact, a company should report any ESG data that is "necessary for an understanding of the development, performance, position and impact of its activity." The first fiscal year in which reports complying with the 2014 Directive have been issued was 2017, after a number of European countries had implemented different reporting criteria, sometimes even stronger than the Directive. For instance, Italy not only implemented the Directive (through D.Lgs. 254/2016) requiring ESG data disclosure as of 2017 for medium and large cap issuers, but also introduced criteria to distinguish the degree of detailed reporting required based on the type of entity. Many EU member States such as France, Netherlands, Denmark, and Sweden had already developed some partial form of mandatory requirement for ESG disclosures prior to the Directive, mainly for state-owned companies. Yet, the 2014 Directive marked a significant expansion (Camilleri, 2017).

In the US, requirements for sustainability reporting are minimal: unlike in Europe, companies' ESG data are only considered material and subject to mandatory reporting if there is a clear financial consideration for them. For instance, the Security and Exchange Commission (SEC) issued in 2010 the SEC Guidance on Climate Change Disclosure according to which such non-financial data should be disclosed if related to a public company's "financial condition, liquidity and capital resources, changes in financial condition and results of operations" (US SEC, 2010). Other examples of specific SEC-regulated ESG issues are board diversity, conflict minerals, and resource extraction. Over the years, efforts increased to get the SEC to adopt widespread mandatory and standardized disclosure requirements related to ESG information (Fisch, 2019). Many large institutional investors, academics, lawyers, and proxy advisors backed a petition to the SEC, calling for rulemaking to define standards for comprehensive corporate disclosure of ESG information; however, the US House of Representatives Financial Services Committee roundly rejected legislative proposals to require widespread ESG disclosure by companies (Temple-West, 2012). Meanwhile, a new bill named the 'ESG Disclosure Simplification Act' (H.R.4329) was introduced by the Financial Services Committee in 2019; if passed by the House, it would require public companies to disclose ESG information in their proxy statements. Hence, nowadays many US companies either issue comprehensive sustainability reports on a voluntary basis, or only incorporate the ESG information deemed material to investors, under existing securities law, in mandated financial reports. NASDAQ's 'ESG Reporting Guide 2.0' (2019), is an example of "support resource for companies" which volunteer to disclose their ESG information.

2.3.2.2 *The Determinants of Sustainability Reporting*

These divergent regulatory trends certainly help explain the different context across the Atlantic. Yet, much of the development of sustainability reporting has been voluntary, and such past tradition also contributes to the explanation. As reconstructed by Ioannou and Serafeim (2017), both in the US and in Europe, voluntary non-financial reporting was first introduced during the 1960s and 1970s, as a result of renewed awareness of the responsibility of business organizations toward society and the environment. In the 1990s, there was a significant growth in the issuance of voluntary corporate sustainability reports in developed markets, due to enhanced pressure and expectation by society for more corporate transparency and accountability. Later on, social and environmental challenges (e.g., poverty, inequality, climate change, resource depletion) renewed pressure on companies by investors, shareholders, and other stakeholders to adopt more systematic approaches to risk management and sustainability reporting. Then, as a consequence of several high-profile corporate scandals and the global financial crisis, a general feeling of distrust grew toward companies' ability to self-regulate. At the same time, investors and information intermediaries in capital markets began to integrate ESG data in their valuation models, creating additional demand for sustainability reporting.

One of the most popular perspectives, *legitimacy theory*, hinges on the notion of a 'social contract' between a business and society. In this view, companies issue sustainability reports "to present a socially responsible image so that they can legitimize their behaviors to their stakeholder groups" (Branco & Rodrigues, 2006, p. 236). For legitimacy theory reporting alleviates societal pressure, since failure to comply with their 'social contract' can possibly lead to sanctions such as less financial capital, fines, and less demand for their products.

Indeed, *stakeholder theory* is another prominent theory to manage the complex and conflicting relationship of companies with their shareholders and external stakeholders. Phillips et al. (2003, p. 481) claim: "Attention to the interests and well-being of those who can assist or hinder the achievement of the organization's objectives is the central admonition of the theory." In this context, governments and regulatory bodies, besides creditors, have also been identified in literature as 'influencers' of CSR activities because companies rely on sustainability reporting as a strategy to address the concerns of their external stakeholders (Roberts, 1992).

These theories should not be seen as competing, but rather complementary. Indeed, most researchers describe stakeholder theory as the dominant and most useful theory in explaining sustainability reporting practice. Also, Tamimi and Sebastianelli (2017) argue that all these theories may be seen as broadly similar, given that they all focus on the different kinds of pressure that stakeholders exert on companies and, above all, on how sustainability reporting is employed by companies to communicate their CSR activities, in order to affect positive perceptions.

To sum up, though governmental regulations can play a role, as pointed out by agency theory and partly by stakeholder theory, many factors need to be considered. Indeed, the theoretical framework around sustainability reporting sheds light on broader motivations and incentives which can lead companies—and have led them for decades before regulations—to disclose their ESG information in a voluntary manner, regardless of legal obligations. Hence, for our case study, these elements dismiss that the presence of regulations mandating sustainability reporting entails an automatic advantage for European companies in terms of disclosure quality and prevents a meaningful comparison between them and US companies. Rather, still within this framework, with specific reference to US public companies and their motivations for voluntary sustainability reporting, Christensen et al. (2019) confirm that firm size is a relevant factor in the US, since greater public scrutiny and relatively lower costs incentivize companies to engage in reporting; also dispersed ownership, with the consequent high information asymmetry, is associated with more efforts in disclosure; interestingly, the correlations found suggest that in sustainability reporting the economic drivers may overlap with those of other traditional, non-CSR voluntary disclosure.

2.3.2.3 Sustainability Performance vs Sustainability Reporting: Emphasis on the Quality of Reporting

The theoretical framework of sustainability reporting regards also its quality. Indeed, sustainability reports are not always credible or effective: For example, stakeholders may fear that information is disclosed only when it is favorable. The voluntary disclosure theory focuses exactly on this, originating from managers' tendency to cleverly release voluntary information on which they have control to enhance organizational economic benefits (Deswanto & Siregar, 2018). Along this view, since sustainability-compliant reputation can raise the present value of future cash flows, corporations tend to increase disclosures of social and environmental information if they get good ESG performance. In contrast, bad-performing companies tend to elude voluntary reporting to reduce costs and avoid being seen in bad light (Clarkson et al., 2008; Dawkins & Fraas, 2011). Thus, for reporting, we must keep in mind the ambiguity of this performance-disclosure relationship.

2.3.2.4 Recent Trends and Standards of Sustainability Reporting

The *Survey of Corporate Responsibility Reporting* (KPMG, 2017) shows that 74% of large and mid-cap companies around the world issued sustainability or integrated reports in 2017, while the share reached 95% among the 250 largest companies worldwide (G250). On a sample of companies in the EU and US leading stock indices sustainability reporting expanded from about 5% in 2006 to 77% in 2015 (Stolowy & Paugam, 2018).

Advances in reporting practices have been favored by the growing diffusion of authoritative guiding principles and standards, such as United Nations Global Compact; OECD Guiding Principles on Business and Human Rights; International Organization for Standardization (ISO); Global Reporting Initiative (GRI); and International Integrated Reporting Council.

GRI is by far the most widely used framework around the world. The share of companies using GRI is 63%, in the global sample of KPMG (2017), while it is 75% in the G250. In the US, as of 2018, 60% of all reporting companies within the S&P 500 adopted the GRI framework (Si2 and Investor Responsibility Research Center Institute—IRRCI, 2018). The diffusion of the GRI should improve the availability of ESG data worldwide: According to the international multi-stakeholder organization promoting GRI since 1997, their standards are "designed to enhance the global comparability and quality of information on [economic, environmental and social] impacts, thereby enabling greater transparency and accountability of organizations" (GRI, 2020, p. 3).

Our analysis below uses GRI standards since their widespread use in Europe and the US allows comparisons on sustainability reporting practices. Besides, GRI's features allow to deepen and detail the quality of sustainability reporting and to grasp differences among users: As highlighted by recent survey data (KPMG, 2017; Si2 and IRRCI, 2018), if reporting companies specify their standard, their adherence to the standard can vary from the mere reference to it, to a more detailed articulation.

While the GRI presents an intricate map of corporate sustainability reporting, its structure also provides room for flexibility by allowing companies to report 'based on' or 'in accordance with' the guidelines. This is also related to the evolutionary process of the GRI: the first official guideline was released in 2000, and updated versions were issued in 2002 (G2 guidelines), in 2006 (G3), in 2011 (G3.1), in 2013 (G4), and in 2016 (Standards). Along this evolution, new elements were added (GRI, 2020); a company's choice to adopt the latest updated version of the standards can signal its commitment to a strengthened reliability and ultimately, to sustainability.

In conclusion, based on trends and standards in reporting practices, noting whether and to what extent listed companies apply GRI standards can give proxy the quality of their sustainability reporting.

2.3.2.5 Hypothesis Development: Sustainability Reporting Practices

By establishing a parallel with the existing literature on credit ratings, Christensen et al. (2019) note that low-quality sustainable reporting can reduce ESG ratings. This risk can be reduced by voluntarily adopting standards as a common baseline emphasizing materiality, thus with less discretion regarding what to report (Kotsantonis & Serafeim, 2019). This is in line with our evidence above that more and more listed companies are applying those standards, namely the GRI.

In accordance with the findings from the literature and practice, we devise the following hypothesis:

Hypothesis 2 Companies with high-quality sustainability reports obtain higher ESG scores.

2.4 Exploring the Gap: Results and Discussion

2.4.1 Methodology

As said, while logically independent, the two hypotheses that we formulated do not exclude each other, and may reinforce or abate each other. Can we explain the gap by a different sectoral mix in EU vs US (Hypothesis no. 1) and/or through different sustainability reporting practices (Hypothesis no. 2) across the Atlantic?

Our empirical analysis relies on ESG ratings expressed as actual numbers.

The ESG ratings of companies broken down into their 'E,' 'S,' and 'G' components and covering a span of four years—extracted from the database Thomson Reuters Eikon—were functional to our analysis. In particular, this data included information on the sector of the companies observed, allowing to verify Hypothesis no. 1. The rest of the data, functional to test Hypothesis no. 2, was hand-collected from the GRI database (https://database.globalreporting.org), picking the necessary information to compute the *GRI score* covering the years 2008–2018—referred to the financial years 2007–2017. The *GRI score* is used to indicate the quality of sustainability reporting, where the database keeps track not only of the GRI-compliant companies, but also of non-GRI reports and missing reports as well. This index thus gives a nuanced picture of the quality of sustainability reporting, beyond the mere acknowledgment of the presence, or lack of, a sustainability report for each company. The use of the *GRI score* is based on D'Apice et al.'s (2020).

The entire list of S&P 500 Index companies was used (505 in total) for the US, while for Europe a group of 468 companies from 22 countries [Austria, Belgium, Cyprus, Czech Republic, Denmark, Finland, France, Germany, Greece, Hungary, Ireland, Italy, Luxembourg, Netherlands, Poland, Portugal, Romania, Slovenia, Spain, Sweden, UK] was drawn from the 610-company list used by D'Apice et al. (2020) focusing on the top-capitalized companies in the EU-28 stock exchanges for which ESG ratings were available in the Thomson Reuters database.

The analysis concerning Hypothesis no. 1 involves mean ESG scores, divided by geographical area (EU vs US), by sector, and by theme (ESG, 'E,' 'S,' 'G'). Mean values of ESG ratings were calculated for the period 2015–2018, instead of referring to a single year, to minimize potential errors or data gaps.

For the analysis on Hypothesis no. 2, as in D'Apice et al. (2020), the *GRI score* was computed in the following way for each year:

- Any company absent from the GRI database was assigned a score of 0.00;
- Any company present in the database, but classified as "non GRI" was given a score of 0.25;
- Any company classified as "citing GRI" was given a score of 0.50;

- Any company classified as "GRI compliant", but below the frontier standard was given 0.75 (over the years, the frontier standard corresponded to GRI3.1 between 2007 and 2013, and to GRI4 between 2014 and 2017);
- Any company classified as "GRI compliant" at the frontier standard was given 1.00.

The GRI data are then aggregated in different ways to highlight results. Furthermore, to show the relationship between the *GRI score* and ESG ratings (articulated in all their themes), correlation coefficients are calculated.

2.4.2 *Results and Discussion*

2.4.2.1 *Hypothesis 1: Sectoral Composition*

As highlighted above, the geographical comparison of ESG ratings is usually based on mean values. In Sect. 2.3, we have seen the relevance of the distribution across sectors/industries. Here we compare the two groups of companies based in the US vs EU, considering their sectoral composition, through mean values.

Table 2.2 shows the synthetic, 'overall'-ESG scores of each sector through the mean of the scores of each company belonging to it. Comparing the two regional groups, it is possible to calculate the geographical gap.

There is indeed an important gap between the US and the EU companies: all sectors combined, weighted by their own share in the respective regional groups, determine a premium for the EU group of 8.06 points, i.e., 14% more than the US group's mean ESG. The gap concerns all sectors but is larger in some—e.g., Energy (24%), Consumer Non-Cyclicals and Industrials (19%)—and narrower in others—e.g., Consumer Non-Cyclicals (8%), Financials (10%), and Healthcare (9%).

To explore the gap in more detail, Table 2.3 shows the breakdown of the ESG rating into its 'E,' 'S,' 'G' elements (themes) and the related geographical differences. This shows quite clearly how the Governance theme is not, on average, responsible for the gap in the overall-ESG ratings: the advantage of EU over US companies in terms of total 'G' score is negligible (0.22 points). Instead, the total 'E' and 'S' scores show even higher geographical gaps than the overall-ESG scores. While the EU

Table 2.2 Mean ESG scores by sector and by geographical area and their gap

Sectors	*Mean ESG EU*	*Mean ESG US*	*Gap in mean ESG*	*Ratio EU/US*
Basic Materials	68.90	62.14	6.76	1.11
Consumer Cyclicals	63.13	52.99	10.14	1.19
Consumer Non-Cyclicals	69.24	64.14	5.10	1.08
Energy	69.90	56.59	13.31	1.24
Financials	61.11	55.75	5.36	1.10
Healthcare	62.59	57.18	5.41	1.09
Industrials	63.16	53.02	10.14	1.19
Technology	63.65	56.13	7.52	1.13
Telecommunications Services	63.34	54.67	8.67	1.16
Utilities	66.89	60.45	6.44	1.11
All Sectors (weighted)	**64.43**	**56.37**	***8.06***	***1.14***

premium can vary according to the sector, the same sectors whose overall-ESG score gap was the largest (Industrials, Energy, Consumer Cyclicals) also show the biggest geographical differences in Environmental and Social performances. Indeed, the case of Utilities shows large gaps both in the 'E' (11.82) and 'S' themes (12.27) but also a large negative gap in 'G' (−10.33): the relatively better 'G' scores attributed to the US over Europe in this sector partly compensate the effects of the other two themes on the average overall-ESG rating. The same applies to Basic Materials.

To sum up, these results suggest that the pattern of higher ESG ratings for Europe holds true in 'E' and 'S' mean scores by all sectors, even if to a varied extent, while the Governance component of ESG ratings does not contribute to the overall EU vs US gap in most sectors. At the same time, the existence of sectoral differences in ESG ratings reinforces the case to account for the sectoral composition of the EU and US groups of companies in the samples used for the comparison, as shown in Table 2.4. Indeed, if the companies forming the two groups (EU and US) are distributed unevenly across sectors, the gap in ESG ratings might also be attributable to this fact, and not only to the possible differences in ESG performance.

Some differences emerge between the two groups. For instance, Basic Materials covers 5% of the US sample, while it reaches 12% in the EU

Table 2.3 Mean 'E,' 'S,' and 'G' scores by sector and geographical area and their gap

Sectors	*Mean 'E' Environment*		*Gap in Mean 'E'*	*Mean 'S' Social*		*Gap in Mean 'S'*	*Mean 'G' Governance*		*Gap in Mean 'G'*
	EU	*US*		*EU*	*US*		*EU*	*US*	
Basic Materials	68.93	59.75	*9.17*	72.03	61.29	*10.74*	63.69	67.58	*−3.89*
Consumer Cyclicals	62.59	49.11	*13.48*	68.74	56.23	*12.51*	55.40	54.27	*1.13*
Consumer Non-Cyclicals	70.54	62.86	*7.68*	72.00	66.61	*5.39*	64.52	63.05	*1.47*
Energy	71.02	53.36	*17.66*	73.46	54.32	*19.14*	63.10	65.21	*−2.12*
Financials	58.11	46.13	*11.99*	62.84	59.18	*3.66*	61.57	60.75	*0.81*
Healthcare	56.93	49.55	*7.37*	68.95	63.60	*5.35*	58.98	56.44	*2.53*
Industrials	60.55	49.58	*10.98*	68.74	54.37	*14.37*	58.28	59.61	*−1.33*
Technology	60.19	50.56	*9.63*	70.92	60.40	*10.52*	57.66	57.49	*0.17*
Telecomm. Services	59.89	52.43	*7.47*	68.11	57.41	*10.70*	56.89	50.89	*6.00*
Utilities	71.43	59.61	*11.82*	68.13	55.87	*12.27*	57.53	67.86	*−10.33*
All Sectors (weighted)	**63.08**	**51.48**	***11.60***	**68.35**	**58.90**	***9.44***	**59.88**	**59.66**	***0.22***

Table 2.4 Sectoral composition of regional groups of companies

Sectors	*No. Companies US*	*Sector Share US (%)*	*No. Companies EU*	*Sector Share EU (%)*
Basic Materials	27	5	54	12
Consumer Cyclicals	79	16	67	14
Consumer Non-Cyclicals	37	7	36	8
Energy	26	5	27	6
Financials	95	19	105	22
Healthcare	59	12	30	6
Industrials	76	15	72	15
Technology	74	15	22	5
Telecommunications Services	4	1	24	5
Utilities	28	6	31	7
All sectors	**505**	**100**	**468**	**100**

group. Financials are the largest sector in both groups, but with different shares: 19% for US vs 22% for EU. Technology and Healthcare are more represented among the US companies than in Europe, accounting for 15 and 12%, respectively, against 6 and 5%. Besides, while Telecommunications Services counts only 1% for American companies, the EU counterpart is 5%.

Table 2.5 shows the effect of a homogeneous composition on the average ESG rating in the context of the EU–US comparison. To achieve a homogeneous distribution between the two groups of companies, new average scores are calculated by weighting the average sectoral ESG ratings of the US companies to the corresponding sector share of EU companies: Hence, a 'hypothetical' US group is designed. This allows a more precise comparison of the ratings, eliminating the differences due to the different distribution across sectors.

By comparing this rating to the original score of the US group of companies, equal to 56.37 (Table 2.2), we note that the hypothetical score obtained through a homogeneous sectoral composition is now only slightly higher by 0.36 or +0.64%.

In conclusion, by applying the same sectoral composition to US and EU companies, there is no significant reduction in the ESG rating gap. Hence, the distribution across sectors of the companies sampled for the

Table 2.5 Calculation of ESG scores for a hypothetical US group of companies

Sectors	*Sector share EU (%)*	*Mean ESG US*
Basic Materials	12	62.14
Consumer Cyclicals	14	52.99
Consumer Non-Cyclicals	8	64.14
Energy	6	56.59
Financials	22	55.75
Healthcare	6	57.18
Industrials	15	53.02
Technology	5	56.13
Telecommunications Services	5	54.67
Utilities	7	60.45
All Sectors (weighted)		**56.73**

EU–US comparison does not explain the identified gap in ESG ratings and Hypothesis no. 1 is not supported.

2.4.2.2 *Hypothesis 2: Sustainability Reporting Practices*

The information collected from the GRI database allows to observe some trends in the sustainability reporting practices of the sampled companies over the years 2008–2018 (referring to fiscal years 2007–2017).

The first two columns of Table 2.6 show the evolution of sustainability reporting practices for US and EU top-capitalized companies by highlighting, for each of the two groups and each year, the share of companies issuing a GRI sustainability report. These results are obtained by extracting from each sample every company obtaining any *GRI score* higher than 0, i.e., regardless of the quality of reporting. Apart from a general progress of sustainability reporting, it can be noted that the advantage of the EU companies over the US (around 20%) remains stable over the years until 2016 [where the data collection for 2017 might be still incomplete at the time we painstakingly collected the data manually from the GRI database].

However, it is necessary to nuance the sustainability reporting behavior by considering its quality. Indeed, as noted above one of the determinants of this reporting can be the legitimation of poor sustainability performance; therefore, the simple fact of issuing a report does not say enough

Table 2.6 Percentage of GRI reporting companies and mean GRI score—years 2007–2017

	% GRI reporting companies		*Mean GRI score*	
	US	*EU*	*US*	*EU*
2007	12	35	0.69	0.69
2008	15	39	0.69	0.69
2009	20	42	0.70	0.70
2010	30	49	0.65	0.68
2011	36	53	0.70	0.75
2012	42	59	0.68	0.74
2013	45	63	0.62	0.67
2014	46	67	0.66	0.74
2015	49	71	0.72	0.75
2016	42	69	0.69	0.74
2017	54	54	0.67	0.76

about the intention to convey one's sustainability/ESG performance in a transparent manner, while the quality of the information delivered can make the difference in this respect. By considering the different *GRI scores* obtained by the companies, the two last columns of Table 2.6, we note that EU companies generally exhibit higher reporting quality. Here, a Transatlantic gap seems to emerge since 2010, when US companies have been distinctly more willing to issue non-GRI reports, less standardized or comparable, and arguably less complete of ESG data.

Company-level GRI scores are then useful to test Hypothesis no. 2, according to which the EU–US gap in ESG ratings can (also) be explained by the fact that companies issuing high-quality sustainability reports obtain higher ESG scores. This hypothesis is verified by calculating the correlation between GRI scores and ESG ratings. This analysis concerns FF.YY. 2015–2017 to focus on the most recent years; an average of the values of these three years was calculated for both numerical variables for each company, in order to minimize potential errors or data.

These results show significant positive correlations between the quality of sustainability reporting (expressed by the *GRI score*) and the ESG rating (Figs. 2.2 and 2.3). These findings confirm the hypothesis with respect to the companies having a high *GRI score*: For example, in the graphs—and more distinctly for the US—the companies with highest-quality sustainability reports (*GRI score* = 1.00) are more concentrated around high ESG ratings (e.g., >40.00).

Moreover, the correlation is stronger for the US than for the EU companies. This finding is expected in light of voluntary disclosure theory, seen above: where there is no reporting obligation, as in the US, those

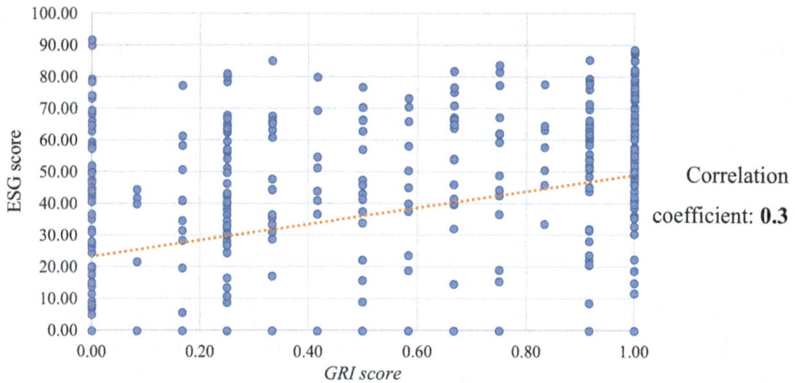

Fig. 2.2 Correlation between quality of sustainability reporting and ESG score, EU companies

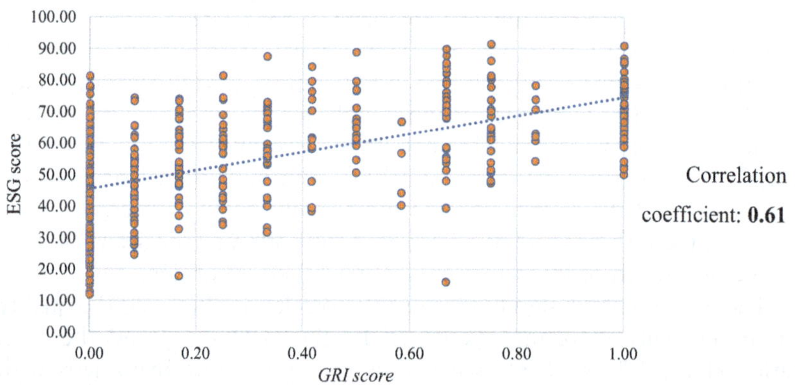

Fig. 2.3 Correlation between quality of sustainability reporting and ESG score, US companies

issuing high-quality sustainability reports are generally the companies with the best ESG performance—ideally represented by the highest ESG ratings—because they can easily be transparent about their operations, releasing more complete and accurate ESG data. Instead, where sustainability reporting is mandatory, such as in the EU, the quality of the report of a company is less likely to reflect its sustainability/ESG performance. Hence, it is not surprising that the *GRI score* is less correlated to the ESG rating for EU companies.

In conclusion, the two findings that, vis-à-vis their US homologues, EU companies more frequently engage in sustainability reporting and have, on average, higher reporting quality does help explain the Transatlantic gap in ESG ratings, supporting Hypothesis no. 2.

2.4.3 Robustness Checks Through Econometric Analysis

The results we obtained through the previous descriptive approach are now put to the test of an important robustness check. Specifically, we run our econometric regressions to verify whether our previous results are confirmed.

Table 2.7 reports the results of OLS regressions based on the data presented above where for each company the dependent variable is either the total ESG rating or the E rating or the S rating or the G rating. Among the regressors, we have included a set of the usual determinants of firm performance. Namely, the set includes the book-to-market ratio (BtM), the operating profitability (OP), the assets growth rate (INV), and the log of the market capitalization (ME). Each specification includes the time, sector, and firm fixed effects, with t-statistics in square brackets and standard errors clustered at firm-level. Nevertheless, given the scope of this chapter, our utmost interest is on three further independent variables: the GRI score (GRI) defined as above, D_EU a dummy variable taking one if the firm is European and zero otherwise, and the interaction variable GRI × D_EU obtained multiplying GRI by D_EU.

Interestingly, three results stand out. First, we confirm the positive and statistically significant effect of the GRI score where companies engaging more effectively in non-financial disclosure obtain higher levels of ESG, E, S, and G. Second, D_EU has a positive and significant coefficient pointing out that EU companies tend to have higher sustainability ratings even after controlling for the usual determinants of firm performance and for time, sector and firm fixed effects. Last, the interaction variable takes a negative value and is generally significant.

By and large, this is in line with the findings of D'Apice et al. (2021) of a stronger role of the GRI score in promoting higher ESG ratings for US-Based holdings vs EU-based holdings, where the fact that non-financial reporting is mandatory for larger sized companies in the EU makes this reporting less informative of a true pro-green corporate approach. On the same line, Clarkson et al. (2013) find "a signaling role for voluntary environmental disclosures. Accordingly, transparent voluntary environmental disclosures increase firm value provided that they are perceived as credible by investors and convey information incremental to what investors already know about the firm's environmental performance. A second potential role for such disclosures is to lower the firm's cost of capital as a consequence of reducing information asymmetry about environmental performance. To serve this role, once again, they have to be viewed as credible and convey incremental information."

2.5 Conclusions

Comparing US companies listed in the S&P500 with a similar group of listed European companies, we have shown that on average EU-listed companies enjoy 14% higher ESG ratings than US companies (64.43 for European companies versus 56.37 for the US ones). Breaking it down, the advantage of European companies is maximum in the E—Environmental component (+22.5%; 63.08 against 51.48), intermediate in the S—Social component (+16.0%; 68.35 against 58.90), and negligible in the G—Governance component (+0.4%; 59.88 against 59.66).

Looking for the determinants of such EU advantage, we could rule out the different sectoral composition of EU companies which, at the most, would give EU companies a minor advantage on US companies. Instead, we could detect a strong EU advantage in terms of quality sustainable reporting. Between 2011 and 2017 almost two thirds of the EU-listed companies included in the sample (62.29%) submitted GRI reports compared to only 44.86% of the US companies. Moreover, the quality of the GRI reports was 8.51% more accurate for the EU vs the US companies. Considering that this larger diffusion of sustainability reporting is also due to the partly mandatory legal set-up, we could envisage that the EU directives were partly at the basis of the ESG advantage of the European companies.

Finally, once accounting for firm performance variables and for sector, time and firm controls, our econometric analysis confirmed both that EU-listed companies enjoy on average higher ESG ratings than US listed companies and that more and better GRI reporting associates with higher ESG ratings for all firms considered in the analysis. However, we also found a minor negative effect. Namely, the positive impact of engaging in more accurate GRI reporting is smaller for EU companies compared to their US counterparts. Thus, the message is clear. Although the partly mandatory nature of the sustainability reporting gives an advantage to EU companies, that advantage has its own limits. Namely, compulsory GRI reporting is less informative of a commitment by companies to engage in the green transition. Therefore, the US companies subscribing to GRI reporting in their constituency, where sustainable reporting is not mandatory, are able to give a stronger signal of their green commitment than it happens for EU companies.

Some caveats are in order. First, we used Asset4 data and, thus, our analysis might not be entirely generalizable to other ESG ratings issued by different providers. Second, the EU vs US advantage might be unstable over time and what we measured up to 2017 could have already changed in more recent years.

As to possible new avenues for future research enticed by our results, we might suggest three areas. First, it would be interesting to distinguish between financial and non-financial companies since financial regulators have been scaling up faster than other regulators to encompass green assets ratios and other ways to consider how the climate crisis builds new types of risks for their supervised entities. Second, the bulk of our analysis used GRI reporting as the only proxy for sustainable reporting. However, even though the GRI continues to be the most widespread sustainability reporting framework, it would be interesting to study whether using other non-financial reporting frameworks to proxy for firms' approach to the green transition would lead to analogous results. Finally, it could be interesting to analyze whether a company's ESG rating tends to increase not only in terms of the quality of its GRI reporting but also if that company subscribes to more non-financial reporting schemes thus possibly showing a higher engagement to the sustainable transition.

Annex

Table 2.7 Panel regression of EGS over GRI score

	(1) ESG	(2) E	(3) S	(4) G
GRI	0.0676***	0.0797***	0.0941***	0.0789***
	[7.1761]	[7.2984]	[8.8934]	[6.0868]
GRI × D_EU	−0.0446***	−0.0693***	−0.0346**	−0.0277
	[−3.4061]	[−4.2838]	[−2.0987]	[−1.4037]
D_EU	0.0273***	0.0429***	0.0311***	0.0188***
	[7.0236]	[8.2243]	[6.3977]	[3.1324]
BtM	0.0143*	0.0173**	0.0281**	0.0177
	[1.9251]	[2.2227]	[2.8717]	[1.2112]
OP	−0.0001**	0.0000	0.0000	0.0000
	[−2.4313]	[−1.0669]	[0.3254]	[−0.6106]
INV	−0.0047*	−0.0129***	−0.0137***	−0.0030
	[−1.739]	[−3.9365]	[−4.402]	[−0.6044]
log(ME)	0.0094*	0.0161**	0.0747***	0.0471***
	[1.7084]	[2.0989]	[12.8401]	[6.0098]
Const	0.2330***	0.0711	−0.7274***	−0.2581*
	[2.6142]	[0.573]	[−7.4418]	[−1.9454]
R^2 adj	0.3710	0.2822	0.1668	0.0474
Obs	5141	4644	5141	5141

The table reports the model's estimated coefficients, where the dependent variable is the overall-ESG score or the single E, S, and G dimensions. The independent variables are instead the GRI score (GRI), a dummy variable taking one if the firm is European and zero otherwise, the book-to-market ratio (BtM), the operating profitability (OP), the assets growth rate (INV), and the log of the market capitalization (ME). Each specification includes the time, sector and firm fixed effects, with t-statistics in square brackets and standard errors clustered at firm-level.
Note *** 1%; ** 5%; * 10%

References

Bannier, C. E., Bofinger, Y., & Rock, B. (2019). *Doing safe by doing good: ESG investing and corporate social responsibility in the U.S. and Europe*. CFS (Center for Financial Studies), Working Paper Series No. 621.

Barclays. (2018). *The case for sustainable bond investing strengthens*. Available at: https://www.investmentbank.barclays.com/content/dam/barclaysmicrosites/ibpublic/documents/our-insights/ESG2/BarclaysIB-ImpactSeries4-ESG-in-credit-5MB.pdf. Accessed on 27 June 2020.

Beloe, S. (2016). *What do ESG ratings actually tell us?* Available at: http://www.whebgroup.com/what-do-esg-ratings-actually-tell-us/. Accessed on 27 June 2020.

Branco, M. C., & Rodrigues, L. L. (2006). Corporate social responsibility and resource-based perspectives. *Journal of Business Ethics, 69*(2), 111–132.

Cai, Y., Jo, H., & Pan, C. (2012). Doing well while doing bad? CSR in controversial industry sectors. *Journal of Business Ethics, 108*, 467–480.

Camilleri, M. A. (2017). *Corporate sustainability*. Springer International Publishing.

Capelle-Blancard, G., & Petit, A. (2015). The weighting of CSR dimensions: One size does not fit all. *Business & Society, 56*(6), 919–943.

Christensen, D., Serafeim, G., & Sikochi, A. (2019). *Why is corporate virtue in the eye of the beholder? The case of ESG ratings*. Working Paper.

Clarkson, P. M., Fang, X., Li, Y., & Richardson, G. (2013). The relevance of environmental disclosures: Are such disclosures incrementally informative? *Journal of Accounting and Public Policy, 32*, 410–431.

Clarkson, P. M., Li, Y., Richardson, G. D., & Vasvari, F. P. (2008). Revisiting the relation between environmental performance and environmental disclosure: An empirical analysis. *Accounting Organization and Society, 33*(4–5), 303–327.

D'Apice, V., Ferri, G., & Lipari, F. (2020). Sustainable disclosure policies and sustainable performance of European listed companies. *Sustainability, 12*(5920).

D'Apice, V., Ferri, G., & Intonti, M. (2021). Sustainable disclosure versus ESG intensity: Is there a cross effect between holding and SRI funds? *Corporate Social Responsibility and Environmental Management, 28*(5), 1496–1510.

Dawkins, C. E., & Fraas, J. W. (2011). Erratum to: Beyond acclamations and excuses: Environmental performance voluntary environmental disclosure and the role of visibility. *Journal of Business Ethics, 97*(1), 244–282.

Deswanto, R. B., & Siregar, S. V. (2018). The associations between environmental disclosures with financial performance, environmental performance, and firm value. *Social Responsibility Journal, 14*(1), 180–193.

Doyle, T. (2018). *Ratings that don't rate: the subjective world of ESG ratings agencies*. Washington, D.C.: American Council for Capital Formation. Available at: https://accfcorpgov.org/wp-content/uploads/2018/07/ACCF_RatingsESGReport.pdf. Accessed on 27 June 2020.

Fisch, J. (2019). Making sustainability disclosure sustainable. *Georgetown Law Journal, 107*, 932–966.

GRI. (2020). *Consolidated set of GRI sustainability reporting standards 2020*. Available at: https://www.globalreporting.org/standards/gri-standards-download-center/consolidated-set-of-gri-standards/?g=9c5ae9db-e973-4679-b887-3feb1e95f423. Accessed 16 July 2020.

Griffin, J., & Mahon, J. (1997). The corporate social performance and corporate financial performance debate: Twenty-five years of incomparable research. *Business & Society, 31*(1), 5–31.

Hill, J. (2020). *Environmental, social, and governance (ESG) investing* (pp. 167–183). Elsevier.

Ioannou, I., & Serafeim, G. (2017). *The consequences of mandatory corporate sustainability reporting*. Harvard Business School Working Paper 11–100.

KBI Global Investors. (2019). *KBI global investors publishes region by region analysis of ESG performance*. Available at: https://www.kbiglobalinvestors.com/kbi-global-investors-publishes-region-by-region-analysis-of-esg-performance/. Accessed on 27 June 2020.

Kotsantonis, S., & Serafeim, G. (2019). Four things no one will tell you about ESG data. *Journal of Applied Corporate Finance, 31*(2), 50–58.

KPMG. (2017). *KPMG survey of corporate responsibility reporting*. Available at: https://assets.kpmg/content/dam/kpmg/xx/pdf/2017/10/kpmg-survey-of-corporate-responsibility-reporting-2017.pdf. Accessed on 15 July 2020.

Ktat, S. (2017). *L'impact des critères E-S-G sur la performance financière des entreprises de secteurs controversés*. PhD thesis Université des Antilles – Faculté de Droit et d'Économie.

LaBella, M., Sullivan, L., Russell, J., & Novikov, D. (2019). *The devil is in the details: The divergence in ESG data and implication for responsible investing*. Available at: https://www.leggmason.com/content/dam/legg-mason/documents/en/insights-and-education/whitepaper/lm-qs-the-devil-is-in-the-details-0919.pdf. Accessed on 18 June 2020.

Morningstar. (2019). *Sustainability Atlas: The Nordics keep the title of global ESG leaders*. Available at: https://www.morningstar.com/insights/2019/10/15/sustainability-atlas. Accessed on 27 June 2020.

NASDAQ. (2019). *ESG reporting guide 2.0–A support resource for companies*. Available at: https://www.nasdaq.com/docs/2019/11/26/2019-ESG-Reporting-Guide.pdf. Accessed on 13 July 2020.

Nitsche, C., & Schröder, M. (2019). *Are SRI funds conventional funds in disguise or do they live up to their name?* ZEW Discussion Paper No. 15-027.

Nordea Markets. (2018). *A multi-billion EUR Nordic opportunity*. Available at: https://nordeamarkets.com/wp-content/uploads/2018/09/ESG_140918.pdf. Accessed on 27 June 2020.

Peiró-Signes, A., & Segarra-Oña, M.-V. (2013). Trends in ESG practices: Differences and similarities across major developed markets. *Sustainability appraisal: Quantitative methods and mathematical techniques for environmental performance evaluation* (pp. 125–140). Springer.

Phillips, R., Freeman, R. E., & Wicks, A. C. (2003). What stakeholder theory is not. *Business Ethics Quarterly, 13*(4), 479–502.

Roberts, R. W. (1992). Determinants of corporate social responsibility disclosure: An application of stakeholder theory. *Accounting, Organizations and Society, 17*(6), 595–612.

Rowley, T., & Berman, S. (2000). A brand new brand of corporate social performance. Business & Society, *39*(4): 397–418.

S&P Global. (2019). *Navigating the ESG risk Atlas*. Available at: https://www.spglobal.com/en/research-insights/articles/navigating-the-esg-risk-atlas. Accessed on 17 July 2020.

Stolowy, H., & Paugam, L. (2018). The expansion of non-financial reporting: An exploratory study. *Accounting and Business Research, 48*(5), 525–548.

SustainAbility. (2010a). *Rate the raters phase one. Look back and current state*. Available at: https://sustainability.com/wp-content/uploads/2016/09/sustain_ability_ratethe_raters1.pdf. Accessed on 17 July 2020.

SustainAbility. (2010b). *Rate the raters phase two. Taking inventory of the ratings universe*. Available at: https://sustainability.com/wp-content/uploads/2016/09/sustain_ability_ratethe_raters2.pdf. Accessed on 17 July 2020.

SustainAbility. (2011). Rate the raters phase three. Uncovering best practices. Available at: https://sustainability.com/wp-content/uploads/2016/09/rtr_phase_3_report.pdf. Accessed on 17 July 2020.

Sustainable Investment Institute (Si2) and Investor Responsibility Research Center Institute (IRRCI). (2018). *State of sustainability & integrated reporting 2018*. Available at: https://www.weinberg.udel.edu/IIRCiResearchDocuments/2018/11/2018-SP-500-Integrated-Reporting-FINAL-November-2018-1.pdf. Accessed on 16 July 2020.

Taliento, M., Favino, C., & Netti, A. (2019). Impact of environmental, social, and governance information on economic performance: Evidence of a corporate "sustainability advantage" from Europe. *Sustainability, 11*(6).

Tamimi, N., & Sebastianelli, R. (2017). Transparency among S & P 500 companies: An analysis of ESG disclosure scores. *Management Decision, 55*(8), 1660–1680.

Temple-West, P. (2012). US Congress rejects European-style ESG reporting standards. *Financial Times*, July 12.

US SEC. (2010). *Interpretation: Commission guidance regarding disclosure related to climate change*. Available at https://www.sec.gov/rules/interp/2010/33-9106.pdf. Accessed on 13 July 2020.

CHAPTER 3

Loan Origination and Monitoring Guidelines: How Do ESG Indicators Affect Firms' Probability of Default?

Egidio Palmieri, Enrico Fioravante Geretto, and Maurizio Polato

3.1 Introduction

The new EBA guidelines aim to strengthen banks' ability to assess customers' creditworthiness, proposing an innovative approach to credit risk management (De Laurentis, 2021; EBA, 2020a). The ability of banks to efficiently allocate capital to cover credit risk and constitute the

E. Palmieri (✉) · E. F. Geretto · M. Polato
Department of Economics and Statistics, University of Udine, Udine, Italy
e-mail: palmieri.egidio@spes.uniud.it

E. F. Geretto
e-mail: enrico.geretto@uniud.it

M. Polato
e-mail: maurizio.polato@uniud.it

P. Wachtel et al. (eds.), *Creating Value and Improving Financial Performance*, Palgrave Macmillan Studies in Banking and Financial Institutions, https://doi.org/10.1007/978-3-031-24876-4_3

conditio sine qua non to achieve long-term equilibrium plays a relevant macroeconomic role for the regulator. A reduced ability of institutions to correctly assess creditworthiness compromises loans granted to the real economy, reinforcing the impact of the opposing phases of the economic cycle[1] (EBA, 2020b). From a microeconomic point of view, the EBA guidelines lay the foundations for a paradigm shift, orienting the credit analysis and management processes toward an anticipatory and proactive model (Tirloni & Antonietti, 2020): a critical novelty is represented by the evaluation of *Environment Social and Governance* (*ESG*) factors in determining firms' creditworthiness. Environmental sustainability is of significant interest to governments, legislators, and supervisory bodies. As part of the *Agenda for Sustainable Development*, the United Nations has identified 17 goals (social and environmental) that the 190 member states have committed to pursuing by 2030. These include the fight against climate change, the spread of responsible consumption and production ideals, improving working conditions, seeking gender equality, and guaranteeing access to clean and renewable energy.[2]

Similarly, in 2018 the European Commission published a document called "*Action Plan: Financing Sustainable Growth*" to facilitate the spread of sustainable finance in the European context. It is divided into three main objectives: (i) convey savings flows toward sustainable forms of investment; (ii) limit the risks deriving from climate change, uncontrolled exploitation of resources, and degradation of ecosystems; and (iii) spreading a financial culture of investments more oriented toward the long term and facilitating the transition process employing regulatory interventions. In the banking sector, the EBA has welcomed the idea of sustainable development and contributed to the debate by publishing a *working paper* to introduce a taxonomy and a unique approach to the issue for the entire banking system and financial institutions (EBA, 2021; Ernst & Young, 2019a; McKinsey, 2020b).[3]

[1] Financial resources are thus withdrawn from the real economy in periods of negative economic cycle and banks, in order to recover lost profitability, are induced to invest in markets with a better risk-return profile.

[2] https://sdgs.un.org/2030agenda.

[3] On 30 June 2021, with the 35th update of Circular 285 of 17 December 2013, the Bank of Italy established that the banks' corporate strategies and decisions consider the objectives of sustainable finance and, in particular, the integration of environmental and social and governance (ESG) factors.

Considering the new paradigm expressed in EBA guidelines, in this chapter, we intend to address the following research questions: *(i) Are ESG factors capable of mitigating and reducing companies' default probability?; (ii) To what extent is ESG performance correlated with a reduction in the likelihood of default?; (iii) Is the risk reduction effect differentiated for companies operating in different geographic areas?*.

To provide an answer to these research questions, a *difference-in-difference analysis model* was applied to a sample of 840 European and American listed companies with annual observations from 2000 to 2021: the data were collected from the *Bloomberg* and *Refinitiv DataStream* platforms. After a preliminary overview of the new regulatory provisions defined by the EBA (Sect. 3.2), the work structure analyzes the main contributions offered by the literature concerning the relationship between ESG factors and the probability of *default* (Sect. 3.3). Then, the methodology and dataset indicated are developed in the context of Sect. 3.4, followed by a presentation of the results obtained (Sect. 3.5). Finally, the contribution ends in Sect. 3.6 with brief reflections and ideas for further refining the research topic.

3.2 The Regulatory Framework: What Has Changed?

The "*Final Report – Guidelines on Loan Origination and Monitoring*" aims to guide institutions toward the definition of credit granting and monitoring procedures based on prudential standards, providing an integrated approach to credit risk management while ensuring the protection of borrowers. The entry into force of the guidelines for new credit concessions took effect on 30 June 2021, while for credits subject to change in the terms and conditions, the application was deferred for one year. As a result, the monitoring system will be fully operational on 30 June 2024.[4]

The guidelines are structured in five main sections: (i) internal governance; (ii) loan granting procedures; (iii) pricing; (iv) evaluation of collaterals; and (v) monitoring framework. Finally, a brief, non-exhaustive

[4] Such adjustment times were granted due to the COVID-19 pandemic. In addition, the increased time available will allow institutions to adapt their internal processes, IT infrastructures, the risk appetite framework, the strategic plan and any planning and monitoring tool involved in the implementation of the aforementioned guidelines.

examination of the news and main points covered in the document is presented, leaving out the aspects beyond this work's scope.[5]

3.2.1 *Internal Governance*

As regards the area of internal governance, it is envisaged that the management body has the task of approving the strategy for credit risk in line with the profiles defined in the context of the *internal capital adequacy assessment process* (ICAAP), of the *internal liquidity adequacy assessment process* (ILAAP) and the objectives expressed in the *risk appetite framework* (RAF): the *risk appetite*; *risk tolerance*; *risk limits*, *risk capacity*, and *early warnings* (Ernst & AIFIRM, 2021; Young, 2019b). In addition, management must also validate the credit application approval process and the monitoring system and promote a credit culture by encouraging the development and acquisition of skills and competencies of the entire bank staff.

Points 56–57 introduce the possibility for institutions to include environmental, social, and governance (ESG) factors: (i) in the assessment of risk appetite; (ii) in defining credit risk management policies; and (iii) as well as in all the other procedures directly or directly involved in the provision of credit, maintaining an integrated perspective with the risk profile. In addition, ESG factors are considered to examine the potential deterioration of the financial performance of clients following environmental, social, and governance events that may affect them, such as, for example:

- Climate change;
- Civil liability for damage caused by failure to comply with mandatory environmental parameters;
- The risks associated with the transition process from traditional energy sources to others with a lower environmental impact; and
- The change in customer preferences concerning new types of products.

Points n. 58–59 deal with "sustainable" credit lines, i.e., a concession subject to compliance with specific environmental parameters (Ernst & Young, 2020). Institutions are thus required to build procedures detailing

[5] For these reasons, no detail will be provided regarding part (iv) reported in the text.

the set of activities or projects deemed worthy of credit from the point of view of environmental sustainability and the procedures for verifying the correct use of these credit lines. In addition, a phase of monitoring the results must follow this type of credit disbursement. It is the responsibility of the institutes to ensure that the applicant has developed an adequate reporting and assessment system of the environmental objectives achievement state. These assessments will be carried out using qualitative metrics and, if possible, quantitative objectives for measuring the level of integration with the assumptions of the strategic plan and the purposes expressed in the context of the risk appetite framework. Unfortunately, at the first entry date into force of the **Loan Originationand Monitoring** (LOM) guidelines, a reference set of qualitative and quantitative indicators helpful in assessing the sustainability of concessions meeting the ESG criteria is not available (EBA, 2020c).

Further new aspects around internal governance are represented by the need to define clear and sufficiently documented credit approval and resolution procedures, identifying a hierarchy of responsibilities and powers that reflect the limits and risk appetite of the institution. The latter must be represented by objectively measurable indicators representative of the characteristics of the loan portfolio: (i) concentration; (ii) diversification objectives; (iii) lines of business; (iv) geographical areas; (v) economic sectors; (vi) products; (vii) credit limits; and (viii) maximum exposures. Credit decisions must also be impartial and free from conflicts of interest.[6]

The LOM guidelines initiate a process of accountability of the business lines by entrusting them with developing adequate internal monitoring systems. The operating units that generated the risk will be required to manage the credit for the entire duration of the loan. The strategic decisions taken upstream bind downstream operations. If any misalignments are generated, the latter will take all the necessary corrective actions to readjust the risk objectives defined in risk planning.

Finally, it should be emphasized that the credit management policies in use must be integrated with a series of "capacity limits" defined regarding a series of budget indicators, such as the debt service coverage ratio, the debt-to-equity ratio, and the cash flow to debt service ratio. In

[6] For this purpose, personnel who have: (i) a personal or professional relationship with the customer are excluded from the credit granting decision; (ii) an economic or other interest, direct or indirect, actual, or potential, financial, or non-financial; and (iii) undue political influence on the customer.

other words, banks are required to establish thresholds for the acceptability of credit applications concerning specific key indicators of an economic-financial nature.[7]

3.2.2 The Procedures for Granting Loans

The procedures for granting loans introduce and define a minimum set of information and processes for calculating customers' creditworthiness, which helps make the credit disbursement process more efficient. In the first place, the institutions in the phase of evaluation of the disbursements to individuals are required to have the following information: (i) purpose of the loan; (ii) customer profession; (iii) ability to repay; (iv) composition of the family unit; (v) financial commitments previously undertaken and related obligations; (vi) regular expenses; and (vii) real and personal guarantees. In the case of loans to businesses, the data to be collected are: (i) purpose of the loan; (ii) income and cash flow; (iii) financial position and commitments; (iv) business model; (v) business plans supplemented by economic and financial projections; (vi) real and personal guarantees; and (vii) specific legal documentation (permits, contracts) (Ernst & Young, 2021).

The guidelines specify how the exclusive creditworthiness of the customer should be assessed as a stand-alone component or regardless of the presence of guarantees. This indicator's rationale is linked to the ultimate purpose of granting contracts, namely an exit strategy that can be followed in the event of *worst-case scenarios*. The cash flow generated by the applicant's ordinary transactions constitutes the primary source of reimbursement, which must be assessed according to the characteristics and conditions of the transaction. A second evolutionary aspect is implementing a forward-looking approach in determining the ability to fulfill the customer's obligations, considering the entire time horizon and the set of factors that could alter or compromise the regular repayment of the debt.[8] For example, in paragraphs 107–108, banks are asked to

[7] EBA (2020a), *Guidelines on Loan Origination and Monitoring*, Annex 1.

[8] The characteristics of debt positions with other institutions are considered in this prospective assessment: the amount to be invested; the value of the principal and interest; the residual duration; interest rates; and the amounts not yet reimbursed and the reimbursement behavior.

evaluate, employing a sensitivity analysis, the change in the creditworthiness of applicants following adverse shocks such as the increase in interest rates for variable rate loans; the reduction of income received; and potential unfavorable changes in the exchange rate between the nominal currency of the loan and the customer's income. More in detail, a list of idiosyncratic and market events is provided that could manifest the ability to affect the customer's creditworthiness significantly and, consequently, compromise the relative ability to fulfill contractual obligations. Respectively, these are the following events:

- Severe contractions in turnover or profit margins;
- Serious operational losses;
- Severe management problems;
- The default of an important trading partner, customer, or supplier;
- Serious damage to reputation;
- Severe reductions in liquidity, changes in funding, or an increase in financial leverage;
- Unfavorable changes in the prices of the goods to which the customer is mainly exposed and exchange rates.
- Severe macroeconomic slowdowns;
- Crisis of the economic sector of reference for the applicant and his customers;
- Increased political, regulatory, and geographic risk; and
- Increase in the cost of financing generated by interest rate increases.

The elements based on which the creditworthiness of companies must be assessed are at least the following: (i) the financial position and credit risk; (ii) the organizational structure, business model, and corporate strategy (possibly formalized in specific plans); (iii) credit scoring or rating; (iv) the financial commitments undertaken toward other institutions; and (v) contractual structure of the loan. As can be seen, the rating, from a final and summary element expressing the relative reliability of the customer, becomes one of a set of factors that the bank must consider in the overall evaluation of the counterparty.

For all categories of companies considered and evaluating a series of innovative indicators on each part of the set currently used, the

EBA regulations establish the opportunity to analyze ESG factors in defining creditworthiness (McKinsey, 2020a).[9] In particular, for micro and small businesses, it is indicated that, although an analysis of the specific customer in detail is preferable, it is still possible to carry out the related findings concerning the portfolio of this class of borrowers. A provision follows that regardless of the size of the company: specifies that if customers are in any case characterized, directly or indirectly, by a more significant risk related to environmental factors, banks should use heat maps to carry out more in-depth analysis of the business model and assess the state of compliance with regulations relating to greenhouse gas emissions and the impact of ESG regulations on the customer's financial position.[10]

3.2.3 Pricing

Pricing must be consistent with *risk appetite*, business strategy, product type, customer creditworthiness, and market conditions. This procedure will be detailed according to the kind of customer: for SMEs, the operation is contextualized regarding the entire portfolio of products requested by the customer, while for large companies, the main focus is the operation itself. The pricing will follow the logic of the markup and will be stratified into the following components: (i) cost of capital; (ii) cost of financing and coverage of specific risks; (iii) administrative costs; (iv) cost of credit risk; and (v) other markups. Any significant transaction lower than the cost, including the related markups, must be reported and justified. Furthermore, it is the task of the institutes to equip themselves with monitoring systems aimed at controlling and integrating information on the levels of risk assumed, the prices applied, and the expected profitability (KPMG, 2019).

Regarding pricing, it should be recalled that the disclosure analysis promoted by the EBA in 2019 resulted in the surprising outcome of how the European banks did not show an adequate correlation between the riskiness of the counterparties and the pricing of the related loan transactions (EBA, 2019).

[9] EBA (2020a), *Guidelines on Loan Origination and Monitoring*, Annex 3.

[10] Heatmaps are a methodology of graphical representation of information based on a color coding.

3.2.4 *Monitoring Framework*

The monitoring framework constitutes the infrastructure that allows the continuous management and control of risk levels, detailing the information down to individual exposures (Deloitte, 2021).[11] To achieve a high degree of efficiency and high timeliness of data, the progressive abandonment of manual procedures in favor of automated algorithms for data collection and processing a credit rating is required.[12] It will be the task of the banks to feed their time series to facilitate the identification of adequate early warnings, also by using external sources for data collection (Ernst & Young, 2019b).[13]

Institutions are required to regularly review the creditworthiness of "at least" medium-sized companies, update credit scoring with variable frequency, and, if necessary, adjust the customer's rating[14] In addition, the institutions must conduct a sensitivity analysis considering the external risk factors capable of altering the customer's repayment capacity and the overall amount of the position. Macroeconomic, sectoral, and idiosyncratic factors are added to these factors.[15]

The monitoring framework must consider exceeding the threshold levels of the early warning indicators.[16] The activation of a trigger involves the insertion of the indicator within a watchlist and the activation of the

[11] Information is required to be reliable, complete, up to date, and timely. Specifically, for credit risk, the database must have the following characteristics: (i) depth and breadth, including the main risk factors; (ii) accuracy, integrity, reliability and timeliness of data; (iii) consistency; and (iv) traceability.

[12] The data of interest for credit risk monitoring are: (i) the payment behavior of customers; (ii) the customer's credit risk; (iii) credit risk by geographical position and by sector; and (iv) write-down of exposures.

[13] In addition to this objective, the monitoring system will have to allow institutions to create a customer file and be able to build an overall view for each individual customer.

[14] The variability is given by the complexity of the operations, the size of the customer, and the risk profile following the principle of proportionality. In the face of clear signs of a deterioration in credit quality, it is necessary to increase the frequency of reviews.

[15] The idiosyncratic factors are, for example: the firm's pricing power, prospects for the reference sector; cost structure; relations with shareholders; management quality; and presence and extent of research and development costs.

[16] These thresholds are defined in strategic planning and in the risk appetite framework, in conjunction with the intervention procedures aimed at managing temporary overruns of the assumed risk levels.

intervention procedures.[17] This process must occur without undue delay, and the designated managers, assessing the extent of the severity of the activation of the indicator, will identify the corrective actions to be undertaken.[18] Furthermore, each decision must be documented and shared with the functions affected by the event (KPMG, 2019). The monitoring framework defined by the EBA is configured in (De Laurentis, 2021):

- Regular review, aimed at the complete reassessment of the borrower's creditworthiness, analyzes the risk profile changes. These transactions take place regularly and are followed by a rating correction.
- Continuous monitoring is carried out daily and verifies early warnings and the trend analysis of credit lines.

3.3 Literature Review

A positive correlation between credit ratings and ESG factors is evident in the literature. The line of empirical studies that analyzed samples of medium and large companies has shown that the management of ESG risks is correlated with an improvement in creditworthiness and a simultaneous decrease in the probability of default. A paper based on a sample of Italian and Spanish companies confirmed this relationship by applying a logistic regression (De Valle et al., 2017). The authors observed that social and governance issues have a more significant impact on increasing creditworthiness, unlike environmental variables, characterized by a low level of statistical significance.[19] In opposition to this last conclusion, a

[17] The following events are considered to be signs of deterioration in the customer's credit standing: (i) adverse macroeconomic events; (ii) known adverse changes in the financial position of borrowers; (iii) a decrease in sales revenues; (iv) a significant reduction in operating margins or profit for the year; (v) a significant delay in the execution of a project or investment; (vi) change in credit risk; (vii) an increase in the cost of capital; (viii) a decrease in activities; (ix) an increase in market volatility; (x) a deterioration in the value of the collateral; (xi) legal actions; (xii) unfavorable credit rating migrations; and (xiii) arrears in payments to the institution of 30 days.

[18] The activation of an early warning indicator leads to an increase in the review frequency and a greater request for information from the client being reported.

[19] The explanation of a lower significance of environmental factors is due to the difficult observability of the impacts of the strategies concerning environmental sustainability. For example, the positive effects of reducing CO2 emissions or the transition of production processes to renewable power sources can be seen in a medium to long-term time frame.

Chinese study, again based on logistic regression, but having as a dependent variable no longer the credit rating but the probability of default of plain vanilla bond issues of Chinese companies, confirmed the statistical significance of the environmental variables, in reducing the likelihood of default (Li et al., 2020).[20]

This latest study also highlighted that ESG factors have a signaling power toward the market, demonstrating the company's ability to enhance human resources and create stable and sustainable networks. As regards the scope of risk management, ESG risk is not only correlated with credit risk but also with operational risk. The validity and statistical significance of the environmental, social, and governance components considered individually are also confirmed in work concerning a dataset of 122 listed companies in the *Bombay 500–BSE500* (Bhattacharya & Sharma, 2019). From the evidence of the model, it is possible to observe that ESG factors have a more significant impact on the creditworthiness of small and medium-sized enterprises, unlike those with large capitalization.[21] The latter already benefit from a low cost of capital, unlike small and medium-sized enterprises, whose limited access to the capital market. A work based on a sample of European and American companies has identified in ESG metrics a more remarkable ability to mitigate credit risk for medium-sized companies (Barth et al., 2021; Kiesel & Lücke, 2019). This evidence, confirmed by a quantile regression, made it possible to identify a *U-Shaped* relationship between ESG and CDS factors.[22] Companies that begin to mitigate ESG risk, not yet benefiting from credibility concerning the management of sustainability issues, will derive minor marginal benefits compared to counterparties with an established environmental and sustainability reputation.

Further research has confirmed through quantitative models the improvement effect of ESG performance on creditworthiness. A study on

[20] The evidence obtained cannot be used to disprove the conclusions of the work of De Valle et al. (2017) as the environmental variable adopted by the Chinese paper is standard energy consumption while the former used a set of three variables: (i) resources used; (ii) CO_2 emissions; and (iii) environmental innovation. Therefore, the two studies are not comparable both in terms of regressors and in terms of dependent variable.

[21] According to the authors, customers and the market particularly appreciate companies that do not limit themselves to respecting the minimum environmental and social requirements set by law but strive to adopt higher standards and contribute to the improvement of society.

[22] In the study, the credit default swap rate is considered as a driver for credit risk.

a large sample of American companies operating internationally (in particular, 27,892 companies analyzed and with a time horizon of 37 years) used the linear and quadratic discriminant methodologies to confirm the positive effect of ESG factors on credit ratings (Michalskia & Low, 2020).[23] The authors suggest that regulatory authorities implement ESG metrics mandatory within the rating techniques. Under this perspective, a recent paper based on a smaller sample (of 565 companies) demonstrates through applying a discriminant function that the inclusion of ESG performance increases the predictive power of rating models (Klein, 2019). A significant contribution is using the Altman Z-Score within the credit rating evaluation model (Altman, 1968). As Bhattacharya highlighted, considering ESG metrics generates a more significant benefit in SMEs, although assessments are made more difficult by the lack of data. The use of the Z-Score allows banks, in the application of the EBA guidelines, to extend the valuation models to smaller companies, ensuring more efficient credit allocation.[24]

The effect of reducing credit risk through the implementation and consideration of ESG factors has territorial characteristics: the positive results are more evident in countries with a high focus on sustainability issues and where stakeholders reward the reduction of ESG risk (Hübel, 2020); furthermore, the cost of capital for ESG firms is lower in countries where investor protection is more excellent (Breuer et al., 2018); finally, it was observed that the credit risk mitigation effect is present for European companies, but not for American ones (Barth et al., 2021).

As shown in the previous chapter, the mandatory nature of non-financial disclosure will impose a "carbon premium" on carbon-inefficient financial institutions. In fact, the existence of this spread presupposes a higher level of riskiness for brown companies. Continuing the discussion

[23] The sample is made up of 17,942 investment grade firms and 9950 speculative grade firms. The years of observation are between 1982 and 2019. As regards the methodological approach, the study suggests the implementation of extremely randomized trees (ERT) and random forest (RF) models.

[24] For the purposes of calculating the Z-Score, it is necessary to collect the following indicators: (i) working capital; (ii) total assets; (iii) retention of profits; (iv) EBIT; (v) market capitalization for listed companies only; (vi) total liabilities; and (vii) turnover.

introduced in the previous chapter, we will investigate whether the probability of corporate default discounts an actual reduction in PD due to improvements in ESG performance. More specifically, the question arises as to whether the existence of a "carbon premium" is really priced within the probability of default observed in the market.

In the context of qualitative studies, a widespread orientation can be observed in favor of considering ESG performance in calculating creditworthiness. For example, an Australian case study showed that ESG rating metrics diverge according to the sector and the business model considered (Stubbs & Rogers, 2015). However, it is noted that the ideal characteristics of a rating methodology based on sustainability principles must have the following three factors: (i) objectivity; (ii) transparency of the method; and (iii) uniformity. Unfortunately, these properties are not yet present on the market, and there is no uniformity of vision on the subject (Henisz & Mcglinch, 2019; Rocca, 2021).[25]

A further literature field has highlighted some critical issues regarding adopting ESG metrics. It was possible to observe the existence of non-uniformity in ESG ratings as a manifestation of an additional risk component that a risk-averse investor must bear (De Santis et al., 2020; Gibson et al., 2021). The main implication is that although the inclusion of ESG metrics raises the expected return and tends to improve the credit rating, companies could sustain a partial increase in the cost of capital as a reward for the higher risks perceived by risk-averse investors. The assessments provided by the rating companies are not uniform and convergent: the existence of a *rater-specific bias* is noted, a phenomenon because of which the evaluator who assigns a particular rating to a company in one of the three dimensions making up the scope ESG, will tend to standardize the remaining two areas to that evaluation (Berg et al., 2019).[26]

[25] According to the authors, the diffusion of multiple evaluation techniques and the lack of convergence is the result of the difficulty of observing ESG drivers.

[26] Assume that the evaluator assigns a specific value to the environmental area, the social and governance scoring will be parameterized to that initial value assigned to the environmental area.

Table 3.1 Composition of the European companies database

Index	*N. Firms*	*Area/country*
EUROSTOXX	50	Eurozone*
BEL 20	20	Belgium
CAC 40	40	France
DAX 30	30	Germany
FTSE 100	100	UK
IBEX	35	Spain
SMI	20	Switzerland
FTSE MIB	40	Italy

Source Own elaboration

3.4 Database and Methodology

The initial dataset (source: Refinitiv Datastream) consists of annual observations of the 505 companies that comprise the Standard & Poor's 500 and 335 companies listed in the main European listings[27] (Table 3.1). To build a model extended to all companies operating in Europe and the United States, it was decided to consider all the companies listed in the markets analyzed. About 15 ESG variables were collected for each of them, and another three related to the issuer's riskiness over 21 years by applying a filter in terms of data continuity (Tables 3.2 and 3.3). The observations of the companies that in the year under consideration had null values in at least two variables that make up the three leading indicators of the overall ESG score or for which the probability of default was not available were also eliminated.

The econometric model did not consider all the summary variables in the ESG field offered by Refinitiv. The subset of variables used is the following:

- ESG_{it}, as a summary indicator of the ESG performance of the *i*-th company at the year of observation *t*. It is based on the weighting of an overall set of 396 variables divided into groups and subgroups as set out in Table 3.2;

[27] The Eurostoxx index includes 50 companies from 11 Eurozone countries: Austria; Belgium; Finland; France; Germany; Ireland; Italy; Luxembourg; Netherlands; Portugal; and Spain.

Table 3.2 Summary of variables and description

Acronym	*Variable name*	*Description*	*Source*
CR	Rating Class	Rating assigned to the issuer at the end of the year under consideration	Refinitiv Datastream
PD	Probability of Default	Probability of default over a time horizon of 4 years from the date of detection	
ZSC	Z-Score	Altman's Z-Score (1968)	
ESG	ESG	Total ESG scoring	
CON	Controversies	Scoring of the number of ESG disputes to which the company is subjected in the year of observation	
SOC	Social	Synthetic scoring of the total 165 variables in the social field considered by Refinitiv	
GOV	Governance	Summary scoring of the total 120 variables in the Governance area considered by Refinitiv	
ENV	Environmental	Synthetic scoring Of the total 111 variables in the Environment area considered by Refinitiv	

Source Own elaboration

- ENV_{it}, as a summary indicator of the subset of environmental variables of the i-th company at the observation year t;
- SOC_{it}, as a summary indicator of the subgroup of social variables of the i-th company at the observation year t;
- GOV_{it}, as a summary indicator of the subset of governance variables of the i-th company at the year of observation t.

This chapter decided not to use the 381 detailed ESG variables, as the model does not consider the differences related to the scope of operations of the individual companies analyzed. In addition, using a comprehensive and capillary set of environmental, social, and governance variables and generating multicollinearity issues would have been incompatible with the chapter's objective.

Table 3.3 Dataset statistical summary

	PD	*ZSC*	*ESG*	*SOC*	*GOV*	*ENV*
USA						
Mean	0.004	3.47	55.04	70.47	63.25	66.41
Variance	0.0004	5.35	50.52	31.93	51.78	43.64
Standard Deviation	0.01997	2.31	7.11	5.65	7.2	6.61
Minimum	0.00001	0.00	677	27.86	5.91	15.89
1st Quartile	0.00009	2.83	41.51	60.26	49.78	55.06
Median	0.00083	5.02	57	71.71	65.5	68.96
3rd Quartile	0.36651	8.98	69.59	81.6	78.53	79.81
Maximum	1.00	14.57	100	100	100	100
EUROPE						
Mean	0.00253	3.54	45.93	39.66	58.64	45.4
Variance	0.00015	12.51	55.39	90.62	39.16	94.95
Standard Deviation	0.01243	3.54	7.44	9.52	6.26	9.74
Minimum	0.00001	1.54	37.19	28.68	21.79	33.33
1st Quartile	0.00012	2.73	47.93	41.09	58.93	47.37
Median	0.00093	4.35	55.37	52.08	66.07	57.89
3rd Quartile	0.31996	5.31	77.27	79.07	86.81	82.81
Maximum	1.00	9.47	100	100	100	100

Source Own elaboration

It is assumed that since December 2015, following the Paris Climate Agreement, environmental issues have enjoyed such significance as affecting the assessment of the probability of default of a company. Therefore, it follows that the PD observed on Bloomberg before that date is considered net of this dimension, unlike the post-agreement ones, which internally express a probability of default already corrected for performance in the social and environmental fields and governance.[28]

The objective is to verify the impact of ESG metrics in the creditworthiness assessment procedures in a sample of companies listed in the main European and American lists. For this purpose, the ***difference in difference analysis*** was used, which helps highlight the changes in the dependent variable following the occurrence of events of an exogenous nature (Wooldridge & Imbens, 2009). In addition, the one-year probability of default was used as a dependent variable as a driver of the

[28] The probability of default estimated by Bloomberg is based on the default likelihood calculated using the Merton distance to default (DD) model. Endogeneity problems with the other variables used are excluded.

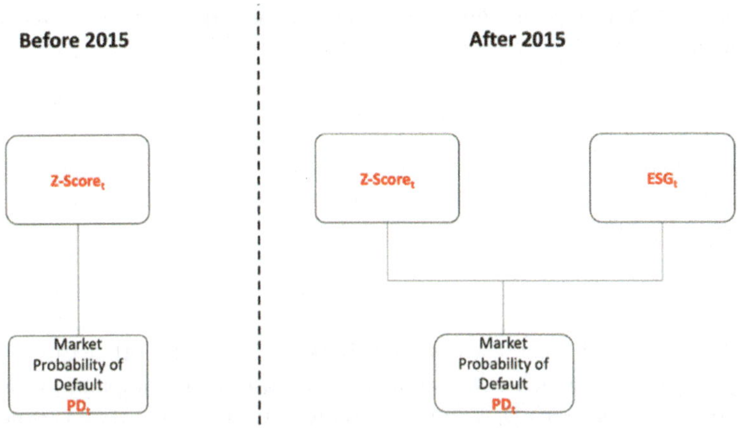

Fig. 3.1 ESG model logical scheme–probability of default (*Source* Own elaboration)

credit risk of the companies considered in the sample (Altman, 1968). The logical model used to build the model is represented in Fig. 3.1. It is noted that the use of the Z-Score is aimed at representing the risk expressed by the balance sheet and income statement items that make up the indicator: (i) working capital; (ii) total assets; (iii) retention of profits; (iv) EBIT; (v) market capitalization; (vi) total liabilities; and (vii) turnover.

Hypothesis 1 ESG factors are negatively correlated with the probability of default. An improvement in ESG scoring causes a decrease in the likelihood of default and a consequent increase in creditworthiness.

Hypothesis 2 The improvement in creditworthiness, resulting from ESG metrics in the evaluation processes, is more remarkable in the sample of European companies than that of American counterparts.

The econometric model used to verify the above hypotheses is the following:

$$\left\{ \begin{array}{r} \mathrm{PD}_{it} = \alpha + \beta_1 * \mathrm{ZSC}_{it} + \beta_2 * \mathrm{ESG}_{it} + \gamma_1 * \mathrm{D}_{\mathrm{time}_{it}} + \gamma_2 * \mathrm{D}_{\mathrm{trm}_{it}} \\ + \lambda_1 * \mathrm{D}_{\mathrm{trm}_{it}} * \mathrm{D}_{\mathrm{time}_{it}} + u_{it} \end{array} \right\}$$

with: (i) PD_{it} = probability of *default;* (ii) α = constant; (iii) ZSC_{it} = Z-Score; (iv) ESG_{it} = summary ESG Score; (v) $\mathrm{D}_{\mathrm{time}_{it}}$ = dummy which

Table 3.4 Difference in difference model framework employed (excluding covariates[29])

	Before ESG (2000–2014)	*After ESG (2015–2020)*	*After–Before*
Control firms	α	$\alpha + \gamma_1$	γ_1
Treatment firms	$\alpha + \gamma_2$	$\alpha + \gamma_1 + \gamma_2 + \lambda_1$	$\gamma_1 + \lambda_1$
Control—treatment	γ_2	$\gamma_2 + \lambda_1$	λ_1

Source Own elaboration

takes a value of 1 for the years after 2014 and 0 in the other cases; (vi) $\mathrm{D}_{\mathrm{trm}_{it}}$ = dummy which takes value 1 in the group of companies treated and 0 in the rest; (vii) $\mathrm{D}_{\mathrm{trm}_{it}} * \mathrm{D}_{\mathrm{tm}_{it}}$ = interaction variable that assumes a value of 1 in the case of companies belonging to the treatment group for the years after 2014; and (viii) u_{it} = error term.

Using this model, it is possible to compare the changes in the probability of default as a function of the risk implicit in the financial statements, the time factor, and the contribution of ESG metrics as an exogenous factor captured by the model. The samples of European and American companies were divided into control groups and treatment groups using a pseudo-random number generation algorithm. All the company observations for each year of observation were included within each group.

Table 3.4 shows the model's operating framework scheme according to the exogenous event and the division into control and treatment groups. γ_1, γ_2, and λ_1 constitute the coefficients of interest for the analysis carried out. They, respectively, measure the coefficient of the effect of time (γ_1), , of the impact of the introduction of ESG measures (γ_2), and the interaction between the two previous variables (λ_1). Hence, the latter constitutes the coefficient of interest.

3.5 Results

The econometric model was applied to the European and American companies' databases, showing a significance of 99.99% in both cases. The

[29] *For the sole purpose of explaining the theoretical framework, the following covariates were excluded: (i) Z-Score; (ii) ESG Score.*

F-test of the American model $F^{USA}{}_{0.001}$ (5; 8.960) = 9.565 is greater than the critical value of the F-test with five regressors with the same confidence interval, $F^{CV}{}_{0.001}$ (5; ∞) = 4.10.[30] The same conclusions apply to the European dataset as $F^{EU}{}_{0.001}$ (5; 3.775) = 26.68 > $F^{CV}{}_{0.001}$ (5; ∞) = 4.10 (Table 2.5).

Consistently with expectations, all the regression coefficients assume a negative sign for both samples. The probability of default over a time horizon of one-year decreases as the Z-Score increases with a confidence of 90% for American companies and over 99.99% for European counterparties. The coefficient λ_1 of the difference-in-difference model assumes a negative value in both datasets with a confidence interval greater than 99.99% for American companies and 90% for European companies (λ_1 = treatment_time). It is concluded that following the management of environmental, social, and governance issues, the probability of default of the companies in the treatment group has a PD at one year on average lower than the control group and those before treatment. Thus, the truthfulness of hypothesis 1 is verified.

The estimate of λ_1 is subject to the variable trend obtained from the multiplication between time and treatment. Both have negative coefficients, although only time is statistically relevant in both samples considered.

It is important to note that an increase in the ESG score contributes to the decrease of the PD at one year by 1.77% for the sample of American companies. In contrast, for the European counterparts, the decline is equal to 6.235% (this factor is indicated as $\theta_{country}$).[31] This percentage improvement is parameterized to the probability of default ($PD_{uncorrected}$), not corrected for ESG factors. For example, a company with a $PD_{uncorrected}$ equal to 10%, following a marginal increase in the ESG Score if it is located in America, will benefit from an expected reduction of 0.177% in the probability of insolvency in the case of a European

[30] CV = Critical Value.

[31] $\Delta PD^{USA} = -0.637 * e^{-5} = -0.01776796624 = -1.777\%$
$\Delta PD^{EU} = -9.253 * e^{-5} = -0.06234622358 = -6.235\%$

Table 3.5 Difference in difference analysis results: USA; EU

Coefficients	*USA*	*EU*
Intercept	0.01499	0.45520***
	(0.08315)	(0.15693)
ZSC	−0.02201	−0.06586***
	(0.01271)	(0.03374)
ESG	−0.01777***	−0.06235***
	(0.01942)	(0.01264)
Time	−0.09305***	−0.22638***
	(0.09253)	(0.06323)
Treated	−1.33246***	−0.07172
	(0.38407)	(0.06960)
Treated_time	−0.03306***	−0.03350·
	(0.03500)	(0.01842)
F-statistic	9.565 on 5 and 8.960 DF	26.68 on 5 and 3.775 DF
*R*2 Adjusted	0.1623	0.1941
P-value	0.000502772	0.00000248

0 "***" 0.001 "**" 0.01 "*" 0.05 "." 0.1 "." 1
Source Own elaboration

company, the reduction will be 0.635 percentage points.[32] These coefficients were estimated with a confidence interval of 99.9% (Tables 2.6 and 3.5).

$$\text{PD}^{\text{country}}_{\text{corrected}} = \text{PD}^{\text{country}}_{\text{uncorrected}} * (1 - \text{ESG}_{\text{score}} * \text{PD}^{\text{country}}_{\text{uncorrected}} * \theta_{\text{country}})$$

$$\theta_{\text{EU}} = 6.235\%; \theta_{\text{USA}} = 1.777\%$$

The model of correction of the probability of default thus proposed causes a distortion effect on the companies whose PD are more consistent (Fig. 3.2). For example, a company with an incorrect PD of 15%, if it reaches an ESG score of 100 points, would benefit from reducing the probability of insolvency, recording a value for the latter indicator of 11.02% in the case of America and 0.97% in Europe. Therefore, banking institutions should impose a maximum limit to reduce the PD depending on the sector and the territorial area to overcome this problem.

[32] All other conditions being equal, ceteris-paribus clause.

Table 3.6 Default probability correction table

	USA			*EU*		
ESG_{Score}	$PD_{uncorrected}$ (%)	ΔPD (%)	$PD_{corrected}$ (%)	$PD_{uncorrected}$ (%)	ΔPD (%)	$PD_{corrected}$ (%)
0	10.00	0.00	10.00	10.00	0.00	10.00
20	10.00	3.54	9.65	10.00	12.47	8.75
40	10.00	7.08	9.29	10.00	24.94	7.51
60	10.00	10.62	8.94	10.00	37.41	6.26
80	10.00	14.16	8.58	10.00	49.88	5.01
100	10.00	17.70	8.23	10.00	62.35	3.77

Source Own elaboration

The adjustment model is proposed below, indicating with $\phi_{\text{sector}}^{\text{country}}$ the maximum applicable reduction[33]:

$$\text{PD}_{\text{corrected}}^{\text{country}} = \text{PD}_{\text{uncorrected}}^{\text{country}} * \left[1 - \text{MIN}\left(\text{ESG}_{\text{score}} * \text{PD}_{\text{uncorrected}}^{\text{country}} * \theta_{\text{country}}; \phi_{\text{sector}}^{\text{country}}\right)\right]$$

The table of the previous example is shown considering a maximum ceiling of $\phi_{\text{sector}}^{\text{country}}$ equal to 25% (Table 3.6).

A significant difference can be observed in the ESG score's ability to reduce the probability of default expressed by the estimators θ_{EU} and θ_{USA}. . The difference between the two indicators is equal to 4.458% and shows a greater tendency toward reducing PDs for European companies compared to their American counterparts. These observations follow the results observed in the sectorial bibliography and verify the truthfulness of hypothesis 2 (Barth et al., 2021; Breuer et al., 2018; Hübel, 2020). The explanation for this phenomenon is due to the propensity of the probability of default to suffer a more substantial reduction following the improvement of ESG performance. In countries where actions to protect the environment are more rewarded and incentivized, the increase in safeguards, social security, and proper governance (Stellner et al., 2015) (Fig. 3.2; Table 3.7).

[33] ϕ Assumes all values between 0 and 100%.

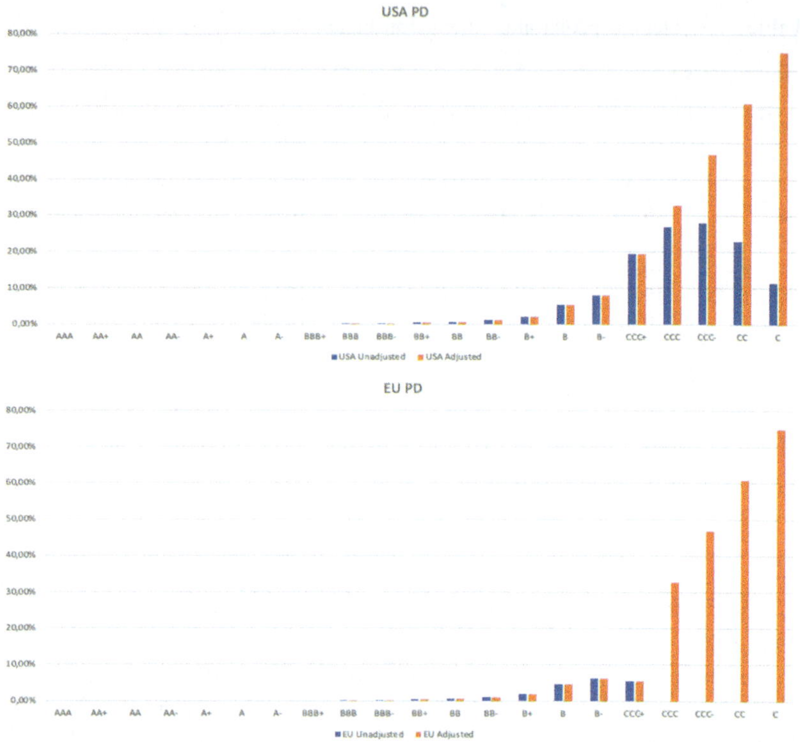

Fig. 3.2 USA and EU PD adjustment compared with ESG Score = 50 and $\phi = 0, 25$ (*Source* Own elaboration)

Table 3.7 Default probability correction table with $\phi = 0.25$

	USA			*EU*		
ESG_{Score}	$PD_{uncorrected}(\%)$	ΔPD $(\%)$	$PD_{corrected}(\%)$	$PD_{uncorrected}(\%)$	ΔPD $(\%)$	$PD_{corrected}(\%)$
0	10.00	0.00	10.00	10.00	0.00	10.00
20	10.00	3.54	9.65	10.00	12.47	8.75
40	10.00	7.08	9.29	10.00	24.94	7.51
60	10.00	10.62	8.94	10.00	25.00	7.5
80	10.00	14.16	8.58	10.00	25.00	7.5
100	10.00	17.70	8.23	10.00	25.00	7.5

Source Own elaboration

3.6 Conclusion

This study analyzed the ability of ESG factors to mitigate issuer risk and reduce the probability of default. Thus, because of the new EBA guidelines on loan origination and monitoring, a quantitative approach has been proposed to correct the probability of default concerning the overall ESG score. In addition, the perspective of external stakeholders, such as credit institutions, was taken, and only publicly available information was used.

The results were obtained by analyzing a sample of 840 companies over a time horizon of 20 years, and the contribution of this work is linked to the measurement of the risk mitigation effect, resulting in a lower probability of default for listed companies in both America and Europe. It was also observed that the ESG Score contributes to reducing PD by 1.777% for American companies and 6.235% for European ones, all other factors being equal. It is noted that the countries that most reward and incentivize the improvement of environmental, social, and governance conditions allow companies to benefit from a more comprehensive effect of reducing the risk of insolvency. It would be interesting to study whether the advantage detected for EU companies vs. US ones is possibly related to the better non-financial disclosure by EU companies stressed in the previous chapter as the primary determinant of building a significant European vantage for EU companies.

A further contribution to the reference banking literature is the introduction of the coefficient $\phi_{\text{sector}}^{\text{country}}$, which allows the proposed model to limit the reduction of the company's probability of default below the level of implicit risk expressed by the Z-Score. From an operational point of view, this is an aspect of particular interest for banks, as, depending on the appetite and risk profile, institutions can define the desired value of $\phi_{\text{sector}}^{\text{country}}$. Moreover, an increase in this coefficient involves a more significant reduction of the estimated probability of default, thus facilitating access to credit in a specific geographical area or sector.

It is noted that some limits characterize this work. Namely, 2015 was defined as the year from which the probabilities of insolvency, observed on Bloomberg, already include the ESG score but the adoption and introduction within companies of ESG policies and strategies followed a process of gradual and non-uniform adaptation for each geographical area and sector. A further criticality is linked to the use within the model of an overall summary indicator for ESG performance: future studies could evaluate the contribution of each specific area to the reduction of the probability of default. Furthermore, within the overall basket of 396 ESG scores available on Refinitiv, further research could construct a set of detailed indicators, according to the sectors to which they belong, capable of perceiving the contribution to the reduction of PD to the change in the basket of variables sectorial relevant.

References

AIFIRM. (2021). *L'integrazione dei fattori ESG nella valutazione del rischio di credito.* Position Paper, 29, 50-56.

Altman, E. I. (1968). Financial ratios, discriminant analysis and the prediction of corporate bankruptcy. *The Journal of Finance, 23*, 594–596.

Barth, F., Hübel, B., & Scholz, H. (2021). *ESG and corporate credit spreads.* SSRN Papers, 2–4.

Berg, F., Koelbel, J. F., & Rigobon, R. (2019). *Aggregate confusion: The divergence of ESG ratings.* MIT Sloan School Working Paper, 5822/19, 28-33.

Bhattacharya, S., & Sharma, D. (2019). Do environment, social and governance performance impact credit ratings: A study from India. *International Journal of Ethics and Systems, 35*(3), 466–484.

Breuer, W., Müller, T., Rosenbach, D., & Salzmann, A. (2018). Corporate social responsibility, investor protection, and cost of equity: A cross-country comparison. *Journal of Banking & Finance*, Elsevier, 96(C), 34-55.

De Laurentis, G. (2021). *Le guidelines EBA su concessione e monitoraggio dei prestiti: Profili critici e implicazioni per banche e debitori.* Bancaria, 70-71.

De Santis, A., Di Marzo, G., Fasano, D., Marlino, M., & Nigro, B. (2020). *Pricing risk-adjusted: il ruolo dei parametri ESG.* PE Working Paper, 20, 16-28.

De Valle, A., Fiandrino, S., & Cantino, V. (2017). The linkage between ESG performance and credit ratings: A firm-level perspective analysis. *International Journal of Business and Management, 12*(9), 53–65.

Deloitte. (2021). *EBA guidelines on loan origination and monitoring, Deloitte,* p. 15. https://www2.deloitte.com/content/dam/Deloitte/rs/Documents/risk/EBA-loan-origination-implementation.pdf

EBA. (2019). *Draft guidelines on loan origination and monitoring.* Consultation Paper, 7.

EBA. (2020a). *Guidelines on loan origination and monitoring.*

EBA. (2020b). *Explanatory note on the EBA comprehensive approach to loan origination.*

EBA. (2020c). *EBA Discussion Paper on management and supervision of ESG risks for credit institutions and investment firms.*

EBA. (2021). *EBA Discussion Paper on Management and supervision of ESG risks for credit institutions and investment firms. Explanatory note on the EBA comprehensive approach to loan origination.*

Ernst & Young. (2019a). *How ESG creates long-term value: Sustainability's new frontier*, 3–5.

Ernst & Young. (2019b). *EBA draft guidelines on loan origination and monitoring—Impact on European banks: Minds made for transforming financial services*, 11–15.

Ernst & Young. (2020). *Risk, returns and resilience: Integrating ESG in the financial sector*, 23–25.

Ernst & Young. (2021). *Loan origination & monitoring: Summary insights.*

Gibson, R., Krueger, P., & Schmidt, P. S. (2021). *ESG rating disagreement and stock returns.* Swiss Finance Institute Research Paper Series, 19/67, 25-27.

Henisz, W. J., & Mcglinch, J. (2019). ESG, material credit events, and credit risk. *Journal of Applied Corporate Finance, 31*(2), 105–117.

Hübel, B. (2020). Do markets value ESG risks in sovereign credit curves? *Quarterly Review of Economics and Finance*, Elsevier, Vol. 85(C), 134–148.

Klein, C. (2019). Quantitative credit-rating models including ESG factors. *The Journal of Environmental Investing, 01*, 14–16.

Kiesel, F., & Lücke, F. (2019). ESG in credit ratings and the impact on financial markets. *Financial Markets, 28*, 263–290.

KPMG. (2019). *EBA draft guidelines on loan origination and monitoring: The ECB and EBA's growing scrutiny on credit standards*, 19–35.

Li, P., Zhou, R., & Xiong, Y. (2020). Can ESG performance affect bond default rate? Evidence from China. *Sustainability, 12*(2954), 1–12.

McKinsey. (2020a). *McKinsey on finance: Perspectives on corporate finance and strategy*,12–20.

McKinsey. (2020b). *A test of resilience: Banking through the crisis, and beyond*, 38–39.

Michalskia, L., & Low, R. K. Y. (2020). *Corporate credit rating feature importance: Does ESG matter?* SSRN Paper, 53–54.

Rocca, E. (2021). *Valenza strategica della gestione dei rischi ESG*. Bancaria, Vol. 4, 59-65.

Stellner, C., Klein, C., & Zwergel, B. (2015). Corporate social responsibility and Eurozone corporate bonds: The moderating role of country sustainability. *Journal of Banking & Finance, 59*, 548.

Stubbs, W., & Rogers, P. (2015). Lifting the veil on environment-social-governance rating methods. *Social Responsibility Journal, 09*, 622–640.

Tirloni, G., & Antonietti, A. (2020). *Le nuove Linee Guida EBA sulla concessione e monitoraggio del credito*, https://www.dirittobancario.it/wp-content/uploads/sites/default/files/allegati/tirloni_g._antonietti_a._le_nuove_linee_guida_eba_sulla_concessione_e_monitoraggio_del_credito_2020.pdf
Wooldridge, J. M., & Imbens, G. W. (2009). Recent developments in the econometrics of program evaluation. *Journal of Economic Literature, 47*(1), 5–86.

CHAPTER 4

Using E from ESG in Systemic Risk Measurement

Ewa Dziwok, Marta Anita Karaś, and Michał Stachura

4.1 How Systemic Risk Affects Financial Institutions

More than ten years after the global financial crisis and dozens of papers about systemic risk, the importance of this risk is no longer in question. It seems that we have also reached a consensus regarding its definition. Generally speaking, systemic risk is "*the risk of a breakdown of an*

E. Dziwok (✉)
University of Economics in Katowice, Katowice, Poland
e-mail: ewa.dziwok@ue.katowice.pl

M. A. Karaś
Wroclaw University of Economics and Business, Wrocław, Poland
e-mail: marta.karas@ue.wroc.pl

M. Stachura
Jan Kochanowski University, Kielce, Poland
e-mail: michal.stachura@ujk.edu.pl

P. Wachtel et al. (eds.), *Creating Value and Improving Financial Performance*, Palgrave Macmillan Studies in Banking and Financial Institutions, https://doi.org/10.1007/978-3-031-24876-4_4

entire system rather than simply the failure of individual parts (...) and denotes the risk of a cascading failure in the financial sector caused by linkages within the financial system, resulting in a severe economic downturn" (CFA Systemic Risk Council, 2022). Alternatively, we may see it as the risk of experiencing a strong endogenous or exogenous systemic event that affects systemically important intermediaries or markets (ECB, 2009, p. 134). Common denominators are the disturbance in the financial systems' continuity and its effect on economic development and societies' well-being.

By 2022, while the COVID-19 pandemic is still ongoing and has lasting effects on economies worldwide, we are moving beyond that classical understanding of systemic risk. The high energy prices in Europe and rising inflation seem to unravel yet another new crisis. Risk exposures that used to be negligible are becoming new systemic risk triggers for financial institutions. Among them, we may count not only financial or economic factors but also geopolitical and environmental ones.

With each crisis, we learn that we must look at systemic risk in new dimensions. It is no longer just the short term and medium term. It seems that more attention should be paid to the long-term perspective. It complicates systemic risk measurement and management, making it even more challenging. The only answer to this challenge is to constantly improve and expand these measures to match developments in the financial and economic systems and to utilize the new data that becomes available. Thus, systemic risk measurement and management are in constant flux.

In light of the newest research and the upcoming standardization of ESG data, we discuss and illustrate how it may become a source of information for systemic risk analysis. We propose a model that augments systemic risk measurement with the environmental factor (E-factor) extracted from ESG data. Markedly, our solution applies to a large set of econometric systemic risk measures. For clarity and transparency, we use the example of the environmental ("E") factor; however, the framework we discuss is universal and can use any of the three factors extracted from ESG scores.

The outline of the chapter is as follows. We start by discussing the role of environmental risk in systemic risk analysis. Then, we discuss how one may extract the environmental risk factor from the ESG score and augment systemic risk measures with it using a beta-independent exposure-based approach. Next, we illustrate and discuss the theoretical properties of our model. Subsequently, we demonstrate its application to the data stylized based on a large sample of systemically important European banks. Our study encompasses the period from 2007 to 2022. Finally, we discuss the model's utility and possible empirical applications by central banks, macroprudential regulators, investors, and other stakeholders of the financial systems.

4.2 Environmental Risk in Systemic Risk Analysis

Much of the existing literature focuses only on climate risk. The conclusions and findings of this literature, referenced in this subchapter, are relevant to our study. However, in systemic risk analysis, one must focus on a wider scope of the environmental impact. Thus, we define environmental risk as the potential for adverse consequences for human or ecological systems that can arise from the impacts of environmental factors, including but not limited to climate change, as well as human responses to such factors (cf. Reisinger et al., 2020).

BIS proposes a simple framework of risk drivers (2021) that translates environmental exposures into financial risk. These risk drivers are the environmental (e.g., climate related) changes that impact economies. They typically occupy one of the two categories: physical risks—the losses related to, e.g., changes in weather, climate, or pollution—that directly impact businesses, institutions, and the economy; transition risks, which arise from the costs of transition toward a low-carbon economy and other sustainable solutions. Notably, "climate risk drivers have a number of distinct features, including unprecedented frequencies, speeds, and intensities and the non-linear form that the risks are expected to take. Together, these factors give rise to a material level of uncertainty as to how climate risk drivers and their impacts will evolve" (BIS, 2021, p. 5).

The Network of Central Banks and Supervisors for Greening the Financial System (NGFS, 2020) illustrates how environmental risk drivers link with financial risks of the banking sector via microprudential and macroprudential transmission channels and their direct and indirect effects. Let us show this in Table 4.1.

Bank of England (2018) also describes the transmission mechanism of environmental risk with physical and transition risk drivers but adds one more category—liability risks. While physical risks refer mainly to materializing of various catastrophic risks, transition risks refer to green

Table 4.1 Transmission channels of environmental risk drivers

	Microprudential channel	*Macroprudential channel*
Definition	The causal chains by which risk drivers affect financial institutions' counterparties, causing financial risk to banks and the financial system	The mechanisms by which climate risk drivers affect macroeconomic factors and how these, in turn, impact banks
Direct effects	The impact on financial institutions' operations and their ability to fund themselves	The impact on macroeconomic indicators, e.g., inflation, labor productivity, economic growth
Indirect effects	The effects on name-specific financial assets held by financial institutions, e.g., bonds, single-name CDSs, equities	The effects on market variables, e.g., interest rates, commodities prices, foreign exchange rates

Source Own elaboration based on NGFS (2020)

innovation-related cash flows. In contrast, liability risks are related to potential compensation payouts that may arise from the above exposures.

It is becoming apparent that transition risks may materialize as unexpectedly high financial losses across financial systems and economies. As Sarah Breeden (2022), the Executive Director for Financial Stability Strategy at the Bank of England, points out, the financial scale of risk may be underestimated when we focus on direct risks of extractive companies, the producers, and sellers of fossil fuels (coal, oil, and gas). In fact, transition risks are also prone to impact assets in other sectors, including petrochemicals, heavy industry, utilities, ground transportation, aviation, shipping, agriculture, and real estate. As Breedeen (2022) argues, "the lost value of these assets is potentially worth trillions or even tens of trillions of dollars".

Lamperti et al. (2021), who study the impact of climate change on global financial stability, reach similar conclusions. The authors show that financial constraints exacerbate climate shocks' effect on the economy, while climate-related monetary damages make financial systems more fragile. Furthermore, their results demonstrate that environmental risks and their cascading multifaceted impacts could increase the frequency of crises by as much as 26—even to 248% (Lamperti et al., 2019).

The recent spikes in inflation across the world, driven, *inter alia*, by increasing brown energy prices, seem to be a new trigger for global systemic risk. According to Professor Robert Engle (2022), this is a manifestation of transition risk that the financial markets should have expected, a market price "tax"[1] on decarbonization. As the investment horizon for brown energy companies decreases, they heavily disinvest (Mehta, 2022). At the same time, the uncertainty regarding their assets' useful life horizon increases, affecting profitability margins and strongly driving prices upwards. According to Engle (2022), this is not a passing trend but rather a new normal; the actual cost of decarbonization—and the prices will remain high, putting strain on businesses, debtors, and financial institutions until the global economy truly decarbonizes.

These observations align with a study by Zhang et al. (2022). They show that environmental changes (especially low-carbon transition) drive the banking sector's risk, climate policy, and banking stability. The study by Tol (2019) indicates that the transition cost may be exceptionally high for developing countries that have the highest social costs of carbon emissions.

Alessi et al. (2022) demonstrate that the potential impact of transition risk on banks' balance sheets is very significant, especially for banks in carbon-intensive economies. They demonstrate that fossil fuel and high-carbon assets may be between 15 and 25% riskier than reflected in banks' risk assessments. In a crisis scenario, this could lead to an increase in losses of up to 40%. Their model shows that fire-sale dynamics, even if triggered by a slight initial depreciation of fossil-fuel or high-carbon assets, lead to significant losses for the whole European Union's banking system and default of many financial institutions (Alessi et al., 2022, pp. 15–19).

Prominent financial institutions and market regulators also admit that environmental factors can significantly influence systemic risk. For instance, the Bank of England (2018, 2021), European Central Bank (2021a, 2021b, 2022), Financial Stability Oversight Council (2021), and International Monetary Fund (2022) point out various environmental risk factors in their systemic risk reports. Similarly, the Financial Stability Oversight Council (2021), Bank of International Settlements (2021), European Systemic Risk Board (2021), European Securities and Markets

[1] A market price correcting mechanism that works as the alternative to the carbon tax that is still not very effective in decarbonizing the global energy markets. Alessi et al. (2022, p. 19) demonstrate that under an orderly transition and actual greening of the economy, banks' transition risk exposure could decrease so significantly that it would reduce the fire-sale losses by a factor of 10 compared to today.

Authority (2022), and European Banking Authority (2022a, 2022b) all recognize that climate change is an emerging threat to financial system stability.

Brunetti et al. (2021) illustrate how significant the shocks related to the degradation of the natural environment and climate change are for economies and financial systems. Similar conclusions are drawn by Toma and Stefanelli (2022), who argue that possessing reliable information about banks' exposure to environmental risk factors will benefit the regulators and the financial industry. Thus, it is vital to research these risks further.

On that note, the European Central Bank (2021a, 2021b) performed an economy-wide European climate stress test. Its results showed how significant the environmental risk may be for systemically important European banks and that climate change "represents a major source of systemic risk, particularly for banks with portfolios concentrated in certain economic sectors and specific geographical areas" (ECB, 2021a, 2021b, p. 3). ECB (2022) is currently running the first climate risk stress test developed to assess the susceptibility of European banks to transition risk. Joint Research Centre of the European Commission studies this issue as well, and their initial results show that many European banks require an additional capital buffer of 0.5% RWA (or 3% of existing capital) to shelter them from systemic risk triggers generated by climate change (Alessi et al., 2022, p. 17).

Despite the unquestioned significance of environmental exposures for systemic risk materialization, methods quantifying them are still scarce. As BIS (2021) states, there is limited data and research reconnoitering how environmental risk drivers feed into the financial risks of banks. Furthermore, it is currently challenging to translate changes in environmental variables into changes in financial institutions' credit, market, liquidity, and operational risk exposures or balance sheet losses (cf. Nieto, 2017).

Toma and Stefanelli (2022) point out that firms do not have sufficient analytical frameworks to combat environmental risk using financial management or internal control tools. Also, policy frameworks designed to deal with environment-related financial risks are bound to be impaired because efficient price discovery is too challenging (Battiston, 2019; Chenet et al., 2021).

Existing econometric systemic risk measures do not explicitly include all environmental risk drivers. They are focused on climate risk. Jung et al. (2021) propose the CRISK model that quantifies the impact of brown emissions-based exposures of banks on their fragility. Authors estimate financial institutions' betas based on "brown investments" and use the SRISK (Brownlees & Engle, 2017) model to incorporate climate risk in systemic risk measurement. A major benefit of this approach is the ability to estimate the individual betas of financial institutions. However, applying this method to Europe requires confidential data, while the output would still consider only a fraction of the actual environmental risk exposure.

Other authors quantifying the link between green finance and larger-scale risk include Battiston et al. (2021), who investigate the spillover of risk in stylized networks, and Sohag et al. (2022), who show that green investments are sensitive to geopolitical risk-based shock transmission. Perhaps the scarcity of methods quantifying environmental risk is related to the fact that econometric methods require precise and granular data that is still very difficult to obtain.

4.3 ESG Data for Systemic Risk Measurement

In recent years, there has been an increasing interest in ESG data reporting and ESG investing. It is coupled with a growing volume of data and intensifying efforts to standardize it. This presents an opportunity to study how environmental, social, and governance issues affect systemic risk and whether the ESG data can be used in systemic risk analysis by financial institutions and market regulators.

So far, ESG factors have been extensively researched in relation to investment (reviews by, e.g., Berg et al., 2022; Billio et al., 2021; Gillan et al., 2021). ESG scores try to capture how investors and companies use ESG factors when running their business (Bahadori et al., 2021; Cornett et al., 2016; Liu et al., 2021), investing (e.g., Bătae et al., 2020; Cormier et al., 2011; Renneboog et al., 2011; Wong & Zhang, 2022) and managing risk (Albuquerque et al., 2019; Boubaker et al., 2020; Bouslah et al., 2018; Kim et al., 2021; Sassen et al., 2016).

Several papers also focus on the relationship between financial institutions' risk and ESG factors. They are either focused on the role of ESG factors in risk management or as risk transmission channels (Brunetti et al., 2021; Candelon et al., 2021; Chiaramonte et al., 2021; Delis et al., 2021; Finger et al., 2018; Gangi et al., 2019; Murè et al., 2021; Neitzert & Petras, 2021; Scatigna et al., 2021). In a most recent study,

Fioravante, Polato, and Palmieri (see Chapter 3) find a significant relationship between ESG ratings and borrowers' probability of default, pointing to a relationship between ESG scores and systemic risk. A handful of papers focus on financial systems' risk (Anginer et al., 2014, 2018; Cerqueti et al., 2021). Although none of these papers proposes econometric methods for measuring systemic risk, they prove that the link between ESG factors and risk exists and is significant.

In a very recent paper, Aevoae et al. (2022) find a statistically significant relationship between two econometric systemic risk measures (Delta CoVaR proposed by Adrian and Brunnermeier (2016) and SRISK proposed by Brownlees and Engle [2017]) and ESG scores. The results obtained for a sample of 367 banks from 47 countries indicate a robust relationship between systemic risk and ESG that is especially strong for the environmental factor (Aevoae et al., 2022, pp. 4, 16–20). Eratalay and Cortés Ángel (2022) draw parallel conclusions from a larger-scale study of blue chip firms, 63 of which are financial institutions.

Mentioned studies show an unused potential and opportunity to use ESG data in systemic risk measurement. To do that, one would have to take an exposure approach that assumes using a readily available environmental score (E-score) as the source of information about financial institution exposure to environmental risks. Such a framework is among the ones recommended by the European Banking Authority (2022a, 2022b). One unquestionable benefit of this approach is cost efficiency—using the data that already exists, that has been gathered by financial institutions for other purposes, and has been pre-processed by external specialized parties.

There are further benefits to using this data. As the Financial Stability Board argues, third-party verification strengthens the reliability of environmental risk data while relying on external metrics available to the broader financial market may "play an important role in avoiding greenwashing risks" (FSB, 2022, *Recommendation II*). Similarly, the *OECD's Report on Environmental Pillar Scoring and Reporting* (Boffo et al., 2020) uncovers that climate risk management and governance are crucial in E-score determination. Because of it, the score can "help investors understand elements of long-term transition" (2020, p. 7) and related longer-term risks.

Also, as EBA (2021) states, the ESG ratings provided by specialized rating agencies account not only for the direct risk exposure to ESG factors but also for the managers' ability to deal with risks and opportunities. This human factor is critical, yet it is difficult to quantify directly in systemic risk measurement. Furthermore, current scoring methodologies "build on a quantitative analysis of key issues identified for each industry (and hence company), as well as qualitative information collected by analysts from public information and engagement with companies" (EBA, 2021, p. 75). Furthermore, score providers compete in the market to provide the best (i.e., most accurate, most transparent, most comprehensive) scores that correlate with effective ESG investment strategies. Thus, it is in their interest to minimize the ESG-washing effects, and they can put most resources into doing this. For these reasons, using ESG scores in systemic risk analysis is potentially very beneficial.

Major developments that should lead to increased availability, transparency, and standardization of Environmental Pillar (Scopes I, II, and III) data are currently taking place. The European Banking Authority (2022a) developed a disclosure template for the ESG factors exposure that large banks will use from January 2023. Moreover, in March 2022, the International Financial Reporting Standards Foundation established the International Sustainability Standards Board. The ISSB has been tasked with the creation of a comprehensive global baseline of sustainability disclosures and is currently working on two new reporting standards: IFRS S1 "General Requirements for Disclosure of Sustainability-related Financial Information" and IFRS S2 "Climate-related Disclosures" (IFRS, 2022).

There seems to be a global need and consensus that these standards are necessary, and the IFRS Foundation is the right institution to provide them. During the 120-day comment period, the ISSB has received more than 1300 comment letters on the two proposals (IFRS, 2022). This may be "a major step toward convergence of the currently fragmented reporting landscape" (KPMG, 2022) that should also help with the problem of green-washing and objectivity of the E-scores.

Even if the ESG data is, to some point, prone to green-washing, no better readily accessible dataset exists that could be used for quantifying the E, S, and G factors in systemic risk analysis. Furthermore, the E-factor is the least subjective, most fact-based, and the least diverse in the way it is calculated by various scorers, suggesting it may serve systemic risk analysis already (cf. Boffo et al., 2020). Finally, the upcoming developments in the

IFRS framework that aim to objectivize sustainability and environmental exposure reporting give grounds to expect further improvements in data quality by the beginning of the year 2023.

4.4 The E-Factor Model

Let SRM be a systemic risk measure (e.g., Marginal Expected Shortfall, Conditional Value at Risk, Systemic Noise Measure, SRISK) that we consider in its absolute or relative version. To add the E-factor to this measure, we follow the rule: the lower the environmental (E) score is, the stronger increase in the SRM is induced (see Eq. 4.2). This property is in line with the findings of the ECB (2022) and the recommendations of the EBA (2021).

To modify the SRM into E-SRM (as we refer to the augmented systemic risk measure) following the above postulate, the empirical time series of the E-factor must be consistent with the SRM series in terms of frequency. Usually, the E-score is published less frequently, so its quotes need to be assigned to the appropriate moments/periods (e.g., days or weeks) of the SRM series. We build the series of the E-factor by extending (e.g., by linear-piecewise interpolation) these sparsely spaced values into the remaining (intermediate) moments/periods for which SRM series values are available. If it is necessary to additionally create E-score values for moments/periods later than the latest quote, one may maintain this latest quote till the end of the considered period. This solution is in line with the findings of behavioral finance theory, which shows how decision-makers utilize the last known data point in their decisions (cf. Kahneman, 2013).

Given the above, without the loss of generality, we assume that both time series are daily. Therefore, $\mathrm{SRM}_{i,t}$ stands for the estimated value of the SRM of the i-th institution on t-th trading day. Likewise, $\mathrm{E}_{i,t}$ is the value of the E-score of the same institution on the same day. Then we define the E-SRM as:

$$\mathrm{E\_SRM}_{i,t} = \mathrm{SRM}_{i,t}\left(1 + \beta\left(100 - \mathrm{E}_{i,t}\right)\right), \tag{4.1}$$

where $\beta > 0$ is a coefficient scaling E-factor influence intensity.

In general, the β coefficient may be time-varying, i.e., $\beta = \beta(t) = \beta_t$ (then β takes the form of a function of time). It may also differ from institution to institution, i.e., $\beta = \beta_i(t) = \beta_{i,t}$. The time variability of the

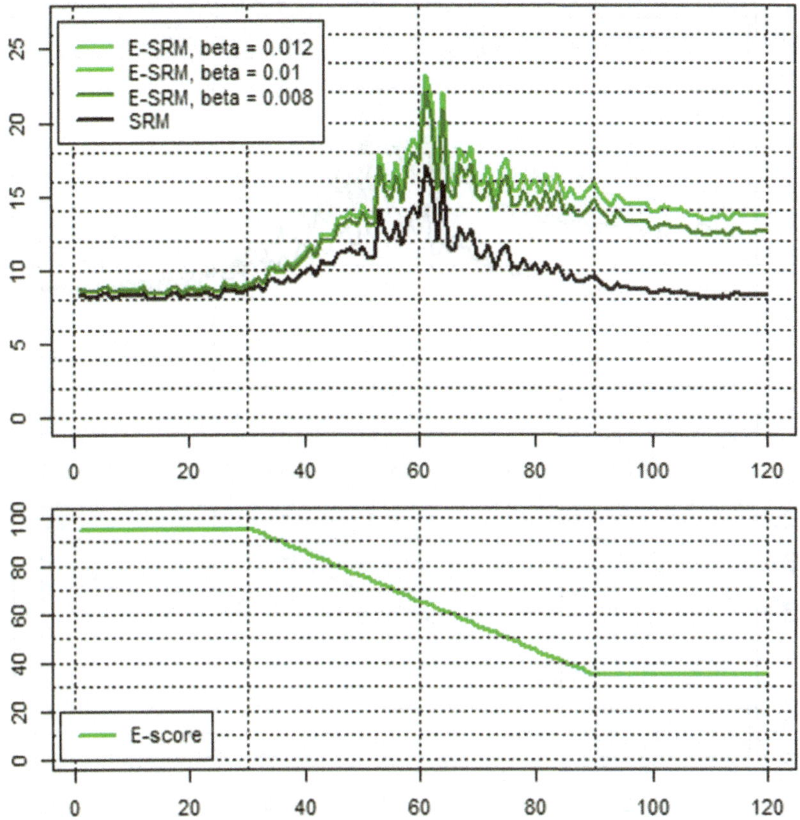

Fig. 4.1 E-SRM–combination A1

β coefficient fits the postulates of the literature, where the time-varying strength of the impact of the E-factor on risks of financial institutions and systemic risk is essential (ECB, 2021a, 2021b, 2022).

Equation (4.1) may be rewritten as:

$$\mathrm{E\_SRM}_{i,t} = \mathrm{SRM}_{i,t} + \beta\left(100 - \mathrm{E}_{i,t}\right)\mathrm{SRM}_{i,t}, \tag{4.2}$$

which demonstrates that the increase in E-SRM is proportional to the decrease in the E-score on a scale defined by the product of β and the current value SRM.

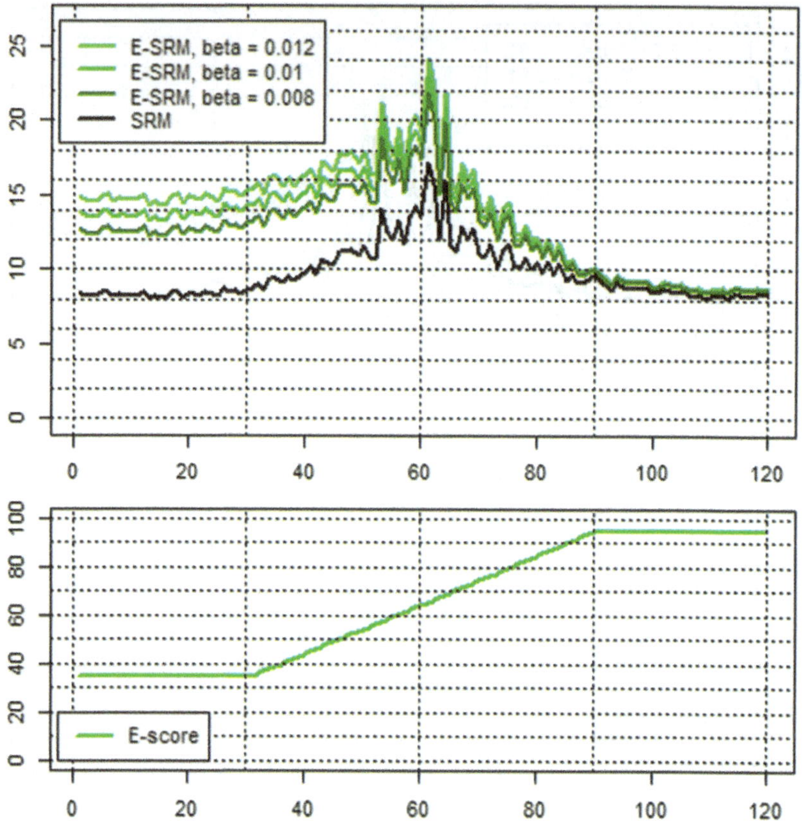

Fig. 4.2 E-SRM–combination A2

In a few selected examples of combinations, we show how significant the influence of the E-factor can be and that the E-SRM course can significantly deviate from the original SRM. For this purpose, we consider two SRM series which represent two different characteristic courses: (A) related to a temporary increase in risk, i.e., a pick (A-shaped), and (B) relatively constant during the period in between two consecutive picks (U-shaped). We combine them with four selected types of E-factor courses that concern only the inner half of the duration: (1) decrease of the E-factor, (2) increase of the E-factor, (3) down-swing and return of the E-factor, and (4) up-swing and return of the E-factor. In each case,

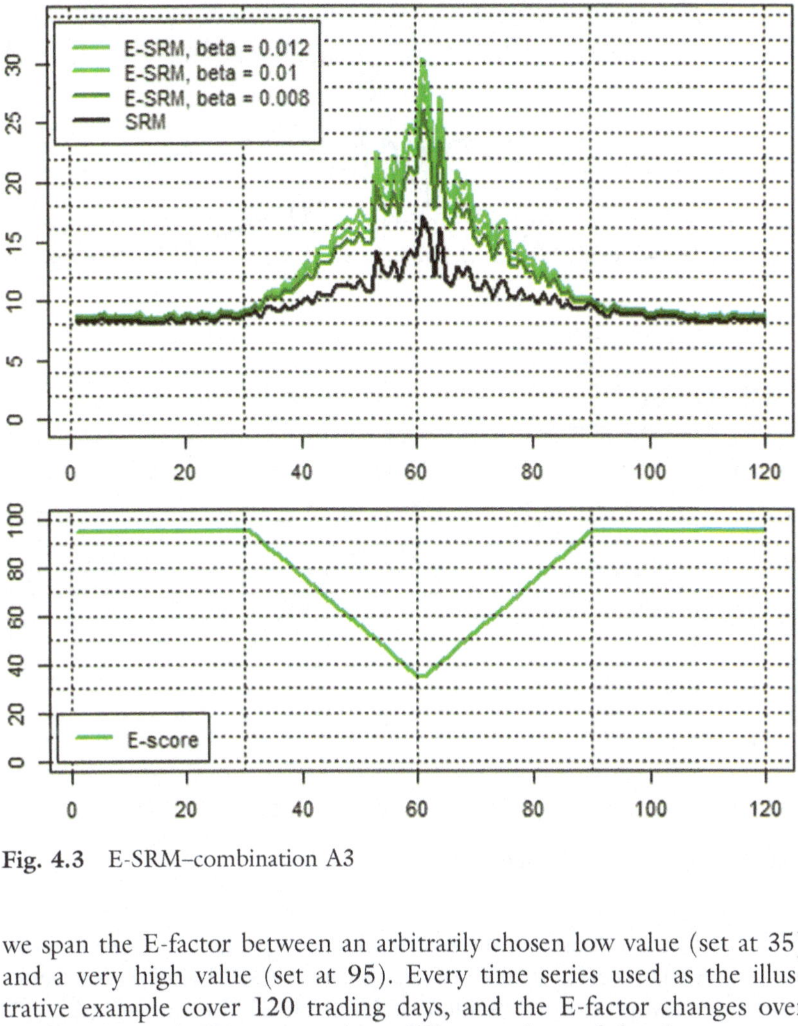

Fig. 4.3 E-SRM–combination A3

we span the E-factor between an arbitrarily chosen low value (set at 35) and a very high value (set at 95). Every time series used as the illustrative example cover 120 trading days, and the E-factor changes over 60 days (31–90). We assume three different values of the time-constant β coefficients: 0.008, 0.010, and 0.012.

Figures 4.1, 4.2, 4.3, 4.4, 4.5, 4.6, 4.7, and 4.8 show how strong the E-score's impact on the SRM can be, especially when the score assumes

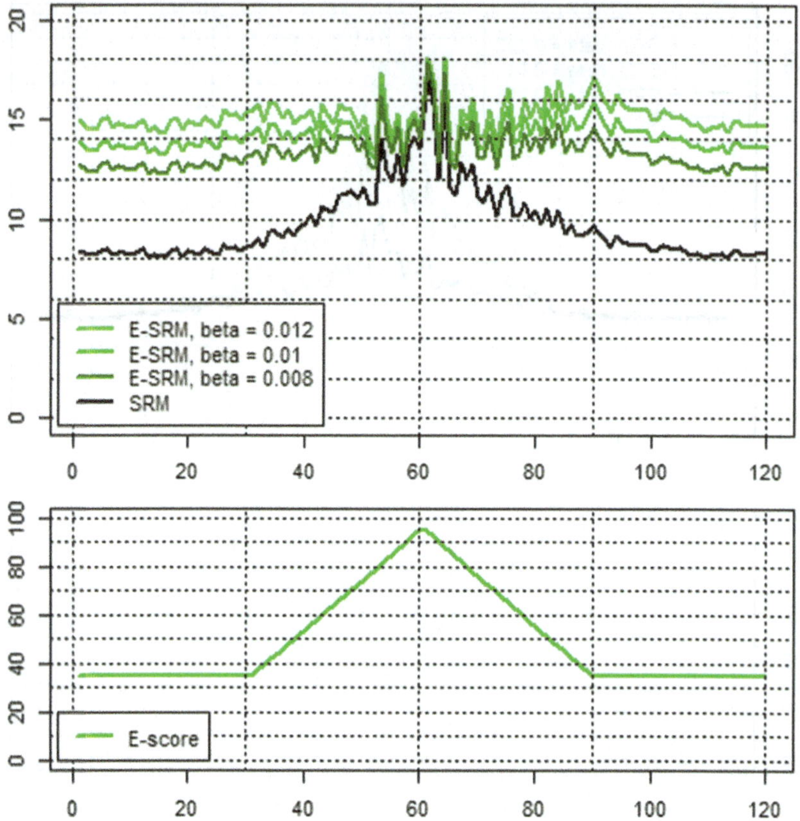

Fig. 4.4 E-SRM–combination A4

particularly low values.[2] Different combinations presented above demonstrate that the meeting in time of changes in the SRM and E-score series may cause a disproportionate increase (A1, A3, B1) or decrease (A2, B2, B4) in the E-SRM compared to the SRM series. It may even change the E-SRM series' general character (A4, B3). Figures 4.3, 4.4, and 4.7 are especially interesting in this context. In the case of combination A3, the E-SRM is particularly strongly amplified, while combination A4 illustrates

[2] Empirical results show that such low E-scores can be traced back to several systemically important European banks (cf. Dziwok et al., 2022).

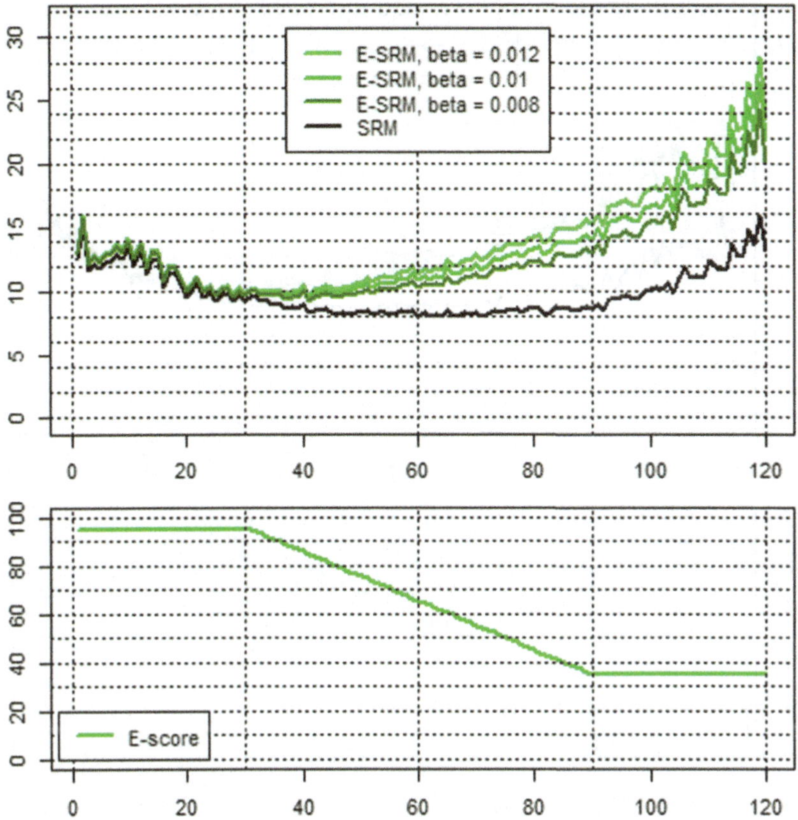

Fig. 4.5 E-SRM–combination B1

the leveling effect of the E-factor. Finally, the B3 combination presents the E-driven peak effect, when the otherwise flat SRM series turns into a peaking E-SRM series after augmentation with the E-factor.

4.5 Examples of the E-SRM Model Application to Stylized Data

In this subchapter, we present four examples that reflect the courses of a prototypical econometric quantile-based systemic risk measure (SRM). It is stylized for systemically important European financial institutions

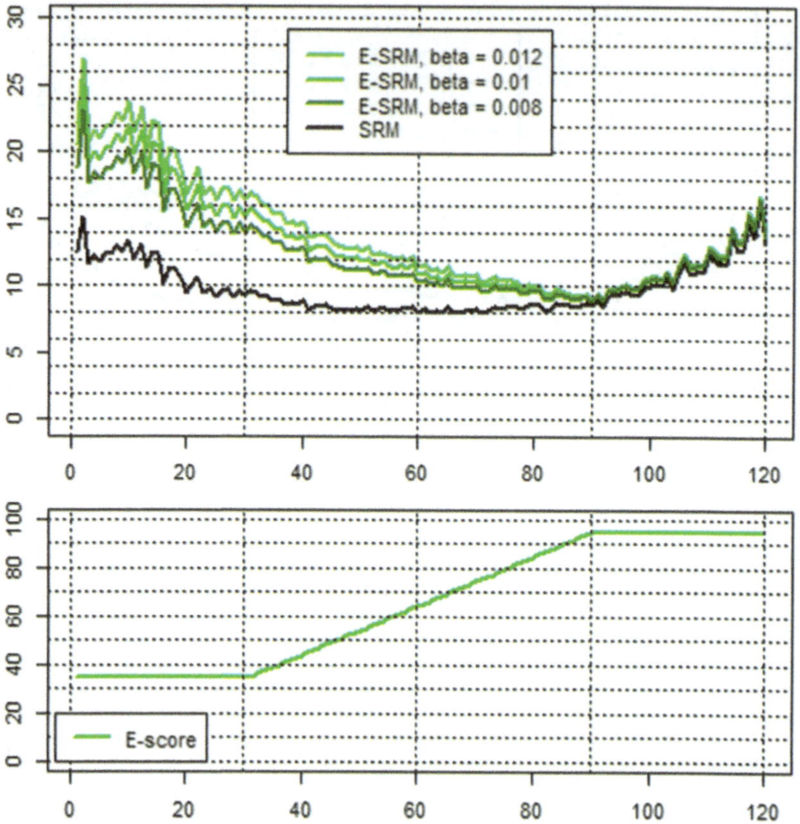

Fig. 4.6 E-SRM–combination B2

grouped by different characteristics. We style the courses of the SRM based on our empirical results from several previous systemic risk studies (Dziwok & Karaś, 2021; Jajuga et al., 2017; Karaś & Szczepaniak, 2020, 2021a, 2021b) and the precise methodology discussed thereof.

In the examples, we use the theoretical SRM measure in its relative form, i.e., SRM%, i.e., as if it was expressed relative to the market capitalization of a given financial institution. In each example, we focus on different periods and geographical locations in Europe, but the common factor for all examples is the materialization of systemic risk measured

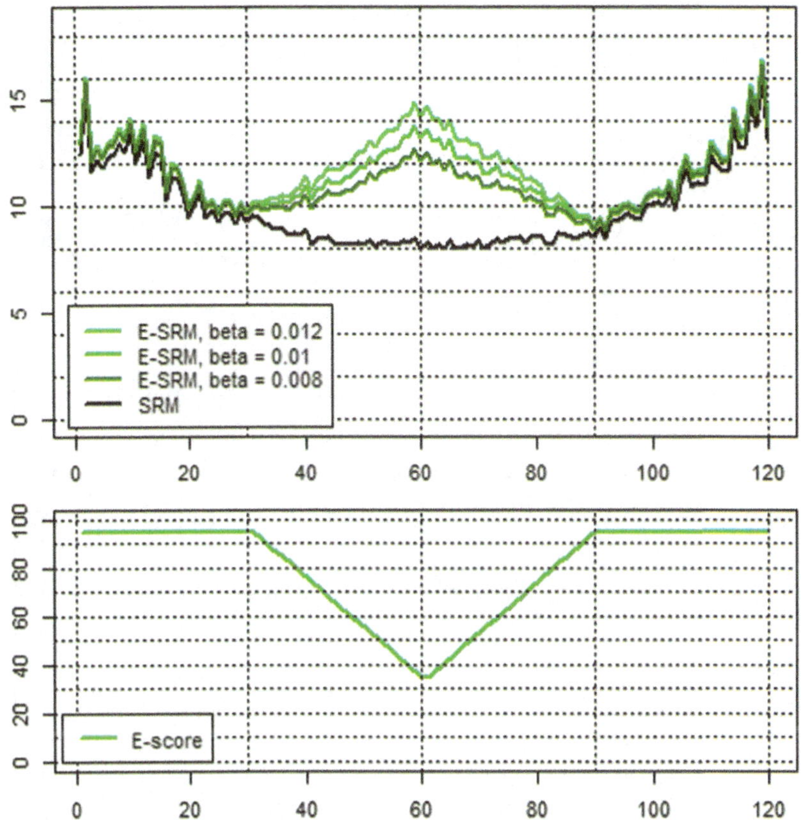

Fig. 4.7 E-SRM–combination B3

for systemically important financial institutions. In selecting systemically important banks, we use the list of Other Systemically Important Institutions prepared yearly by the EBA (2022b).

In each example, we establish a stylized course of the daily SRM measure illustrating the properties described by our previous empirical findings for the period between 2006 and 2022. Then, we attach selected possible courses of the E-factor based on the empirical observation that the E-scores tend either to fall or to stop rising around financially turbulent periods (cf. Dziwok et al., 2022). All so-obtained variants are presented in Figs. 4.9, 4.10, 4.11, and 4.12.

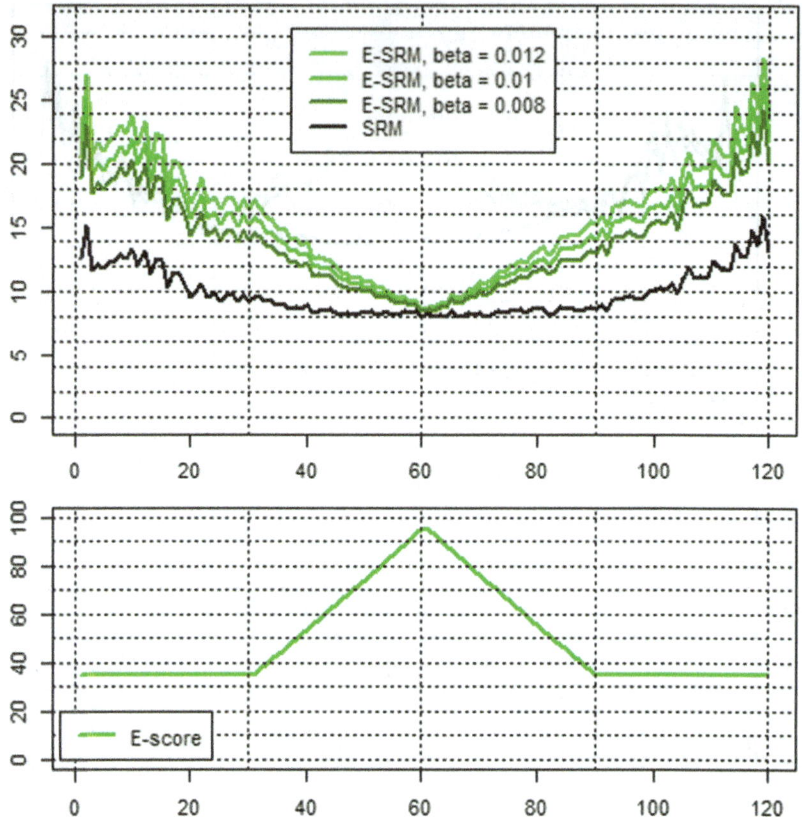

Fig. 4.8 E-SRM–combination B4

Example 1 The first example depicts a typical course of a systemic risk measure for the Nordic and Baltic systemically important banks. Figure 4.9 shows the course of the SRM throughout the study period. Figure 4.10 illustrates (magnifies) the reaction to the global financial crisis that was the most significant manifestation of systemic risk in these countries.

This example depicts several properties of the baseline SRM (the black line) and the E-SRM augmented by the E-factor (the green line). Above all, for the Nordic-Baltic region, the level of financial stability characteristic of systemically important banks is generally very high. Throughout

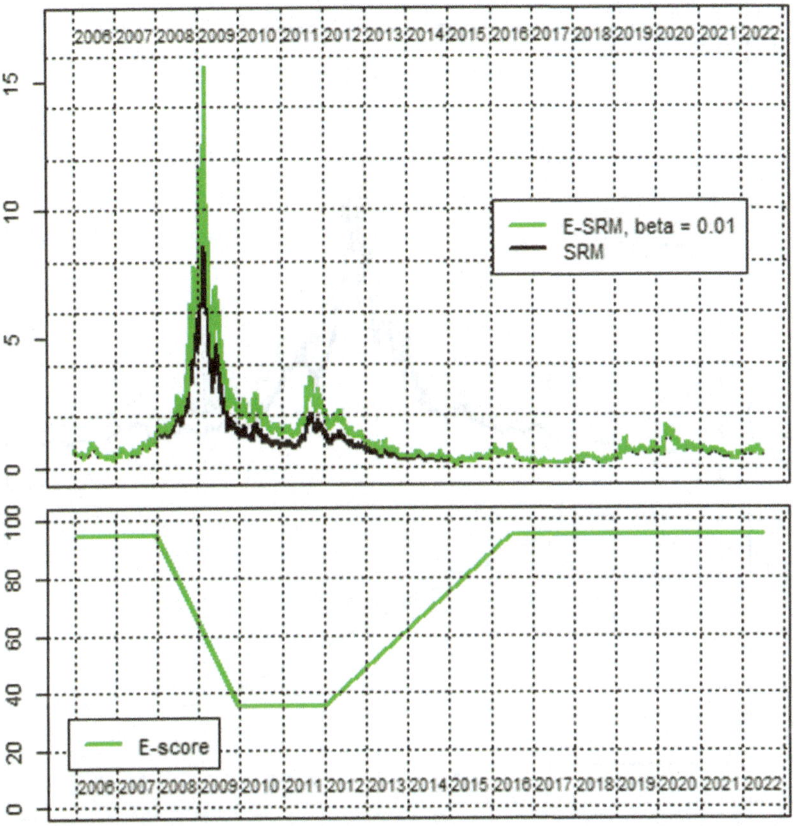

Fig. 4.9 SRM and E-SRM of the Nordic and Baltic OSIIS 2006–2022

the studied fifteen-year period, the SRM measure remains mostly around 0 and 5%, pointing to low systemic fragility. However, during the global financial crisis, there was a large spike in risk that subsided quite fast—over one year.

Figure 4.10 demonstrates how the falling E-factor (bank's increasing exposure to environmental risk) may increase the scale of systemic risk materialization. A crucial property of the model is that by construction, the impact of this exposure automatically increases with the rising levels of the SRM. As discussed in previous sections, this is theoretically justified and empirically expected. In this example, although the assumed beta

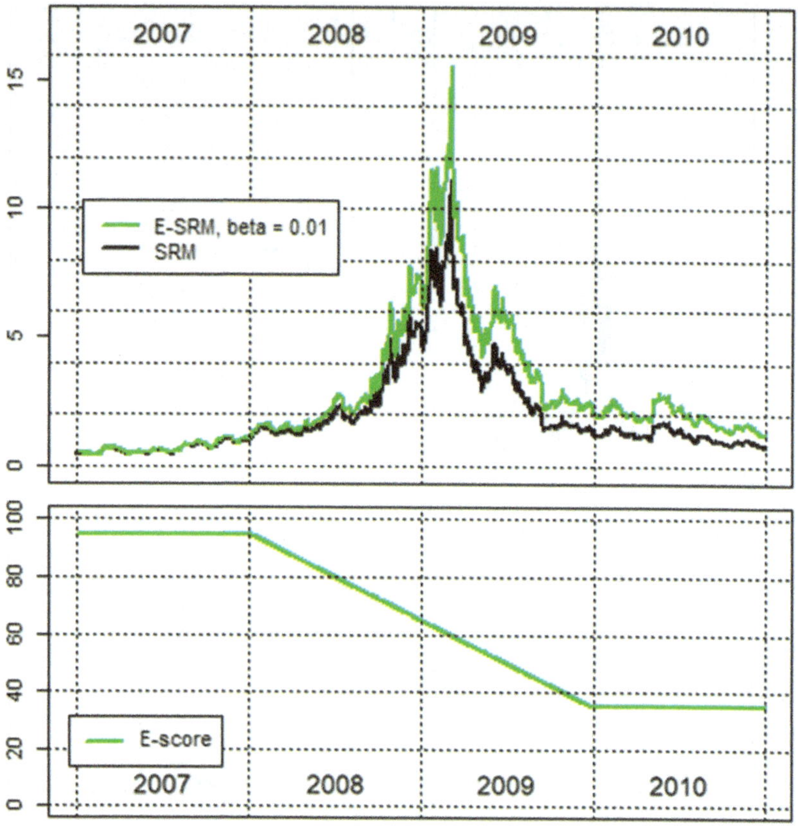

Fig. 4.10 SRM and E-SRM of the Nordic and Baltic OSIIs 2007–2010

coefficient is particularly small (1%), the impact of the E-factor on the SRM reaches 5%—almost one-third of the total systemic risk in its peak of early 2009, when the primary wave of the global financial crisis hit Northern Europe.

Example 2 Systemically important banks of several European countries were particularly strongly hit by the public debt crisis between 2010 and 2013 when the markets reacted to the uneven risk of the sovereign bonds that was reflected by the increasing CDS spreads between euro-denominated bonds of those countries. Banks' exposures were not equal

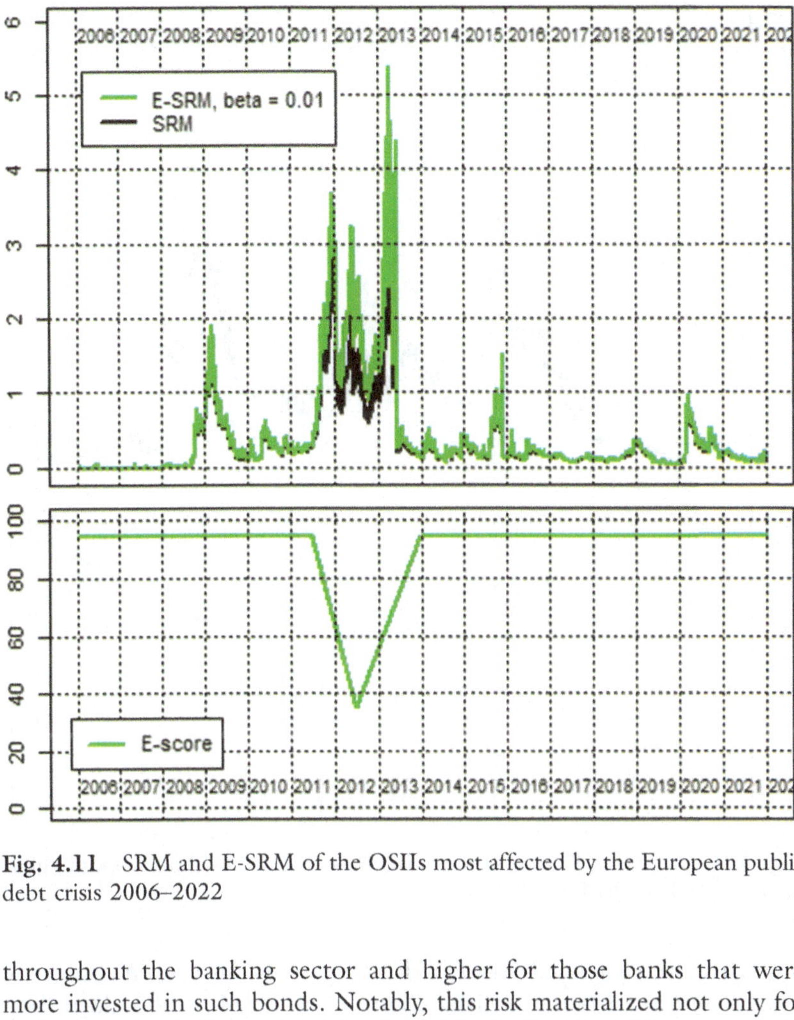

Fig. 4.11 SRM and E-SRM of the OSIIs most affected by the European public debt crisis 2006–2022

throughout the banking sector and higher for those banks that were more invested in such bonds. Notably, this risk materialized not only for such countries as Greece but also for others in Southern Europe and the Balkans. For most affected banks, the systemic risk reaction was sequential to the CDS markets' reactions and lagged by several months.

In this example, we have selected these OSIIs for which the systemic materialization of the European public debt crisis was stronger than that of the global financial crisis. As illustrated in Figs. 4.11 and 4.12, for these banks, systemic risk spikes have a U-shaped recurring pattern and

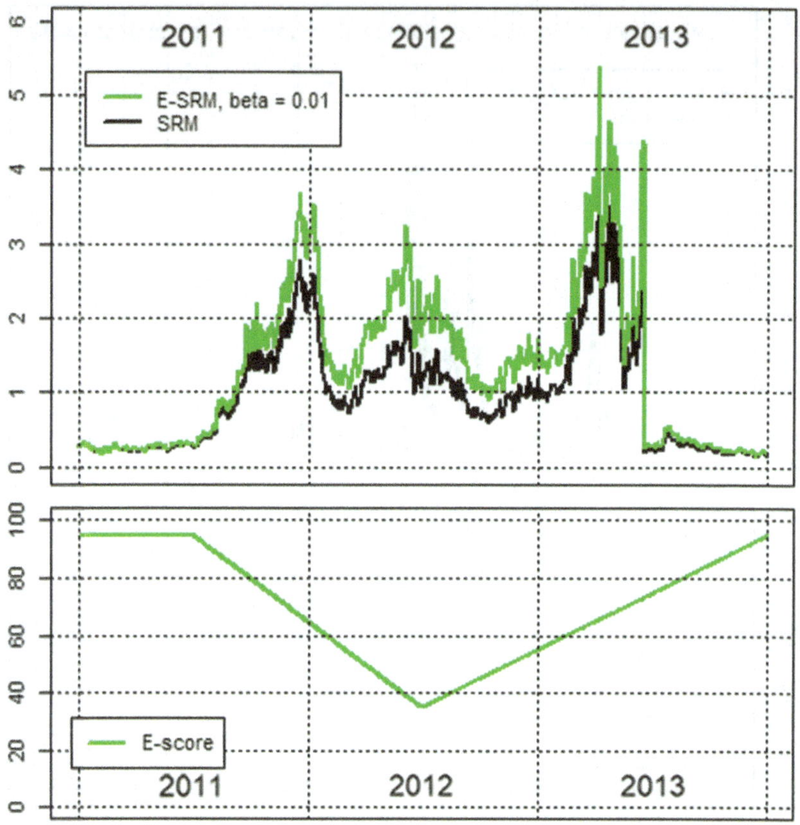

Fig. 4.12 SRM and E-SRM of the OSIIs most affected by the European public debt crisis 2011–2013

a much more prolonged impact than in the previous case. A distinctive property captured by the stylized course of the SRM is the sudden drop in risk that coincides with various rescue measures. They include emergency assets programs and bail-outs, but also mergers and take-overs of the straggling banks. As for the E-SRM, we stylize the E-factor for a temporary drop that may hypothetically be the effect of the government stopping subsidies and tax-relief programs that stimulate green innovation and decarbonization. The impact on systemic risk is significant but smaller than in the previous case.

Example 3 For another example, we have selected systemically important banks of highly industrialized countries, such as Germany, Italy, or France. The courses of systemic risk measures for such banks are characterized by a strong reaction to the global financial crisis that is followed by a more permanent upward shift in the mean level of systemic risk and equally sizable risk spikes in all the subsequent periods of systemic risk materialization (Fig. 4.13).

In this example, it is worth noticing the spike in risk between 2015 and 2017 that corresponds to the low profitability of systemic European

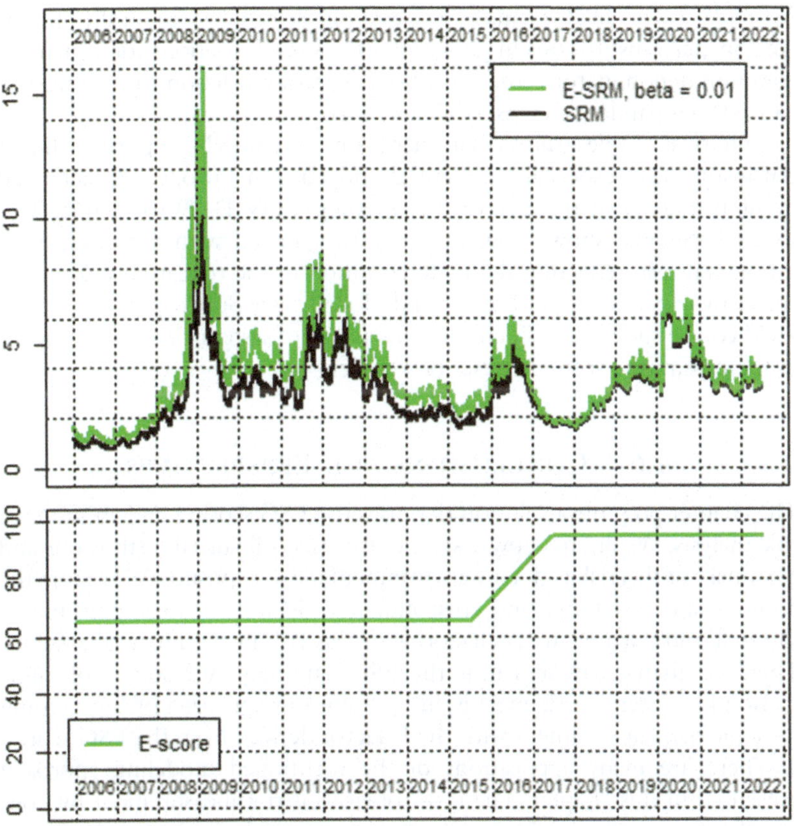

Fig. 4.13 SRM and E-SRM of the OSIIs in highly industrialized European countries 2006–2022

banks imposed by the negative interest rates in the Eurozone. In our simulation, this systemic risk event coincides with the gradual but significant increase of the E-score that may be associated with the fact that the environment of negative interest rates may actually be inductive of investment in green innovation and may lead to decreases in environmental risk exposures. Figure 4.14 demonstrates how the proposed model accounts for this scenario, when the green time series (E-SRM) closes to the black (baseline SRM) time series in 2017. When the hypothetical E-factor increases to 95%, the impact of the environmental risk on the E-SRM decreases to a minimum.

Example 4 The final example is based on selected systemically important financial institutions in geographically varied locations for which the common denominator is the exceptionally strong reaction to the ongoing COVID-19 pandemic (Figs. 4.15 and 4.16).

Empirical studies show that there are systemically important financial institutions, especially in Central and Eastern Europe, characterized by high fragility in the face of the ongoing COVID-19 pandemic. This stylized example shows how this may be coupled with increased environmental risk, reflecting the materialization of the transition risk we are currently experiencing in Europe and globally (see Subchapter 2). The E-SRM course shows how significant this aspect of systemic risk may become in the pessimistic scenario described earlier.

4.6 Conclusions and Perspectives

The chapter presents a solution that uses the ESG scoring data in systemic risk analysis. We discuss how systemic risk affects financial institutions and what part of this risk is due to environmental risk exposure. We report the findings about systemic and environmental risk interactions, pointing to the past and current materializations of this risk in the financial systems. Then we discuss why and how the ESG data may be a source of information for systemic risk analysis and present our approach of augmenting systemic risk measurement with the E-factor derived from the ESG scores.

There are many applications of the mentioned modeling approach presented in this chapter. The most obvious is risk measurement by policymakers, e.g., central banks or macroprudential regulators. However, entities exposed to systemic and environmental risk, like financial institutions, can also use such a method to measure their changing exposure

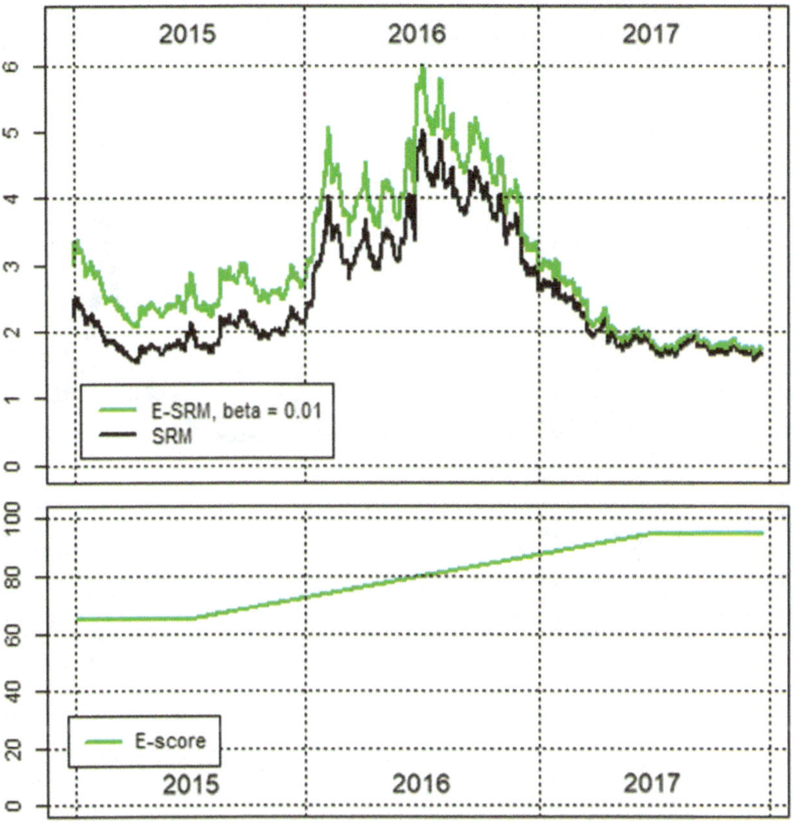

Fig. 4.14 SRM and E-SRM of the OSIIs in highly industrialized European countries 2015–2017

to these risks. Furthermore, other stakeholders, such as investors and debtors—businesses and individual clients of banks—can utilize such a tool when choosing a bank. It might be especially worthwhile when making longer-term financing decisions.

A less obvious but potentially even more valuable application of the proposed model is its use in scenario analyses and stress tests to understand how different decisions related to the size of the E-factor could affect banks' systemic risk exposures. Similarly, the model can be used to analyze and stress-test systemic risk exposure of each financial institution

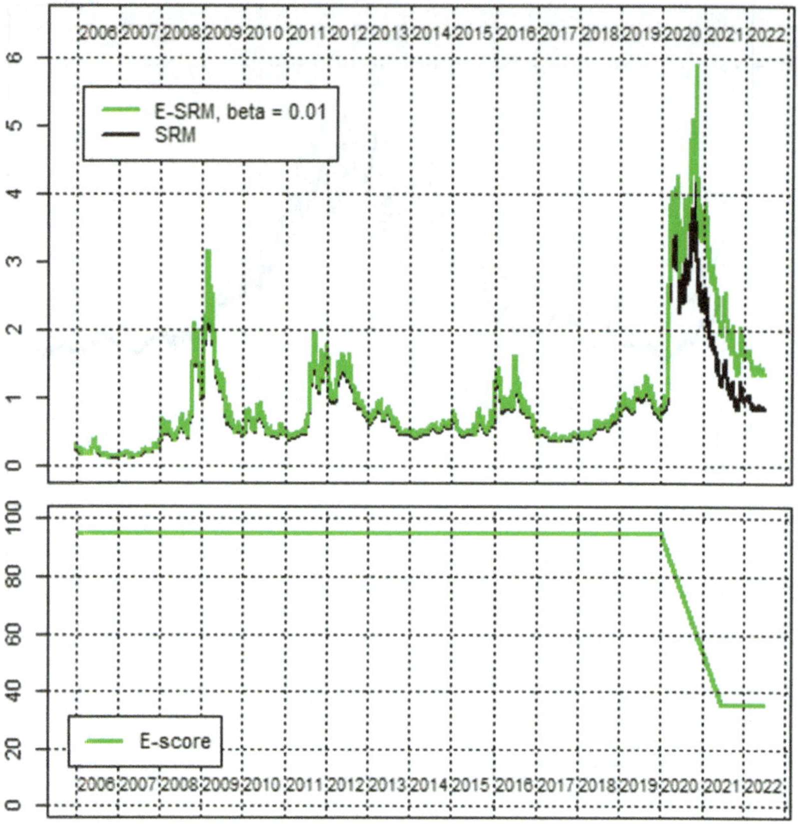

Fig. 4.15 SRM and E-SRM of the OSIIs most affected by the COVID-19 pandemic 2006–2022

in the context of the changing sensitivity to environmental risk (changing betas). Consequently, each of the two variables, the E-factor and the beta, can be stressed—separately or in combination.

The simplistic and transparent construction of the model makes it applicable to a broad spectrum of network-based stress-testing analyses performed by macroprudential regulators and central banks in financial stability analyses. In this context, augmentation can be performed not only on individual banks but also on banking networks, where the effect of the spillover of risk may be observed and measured.

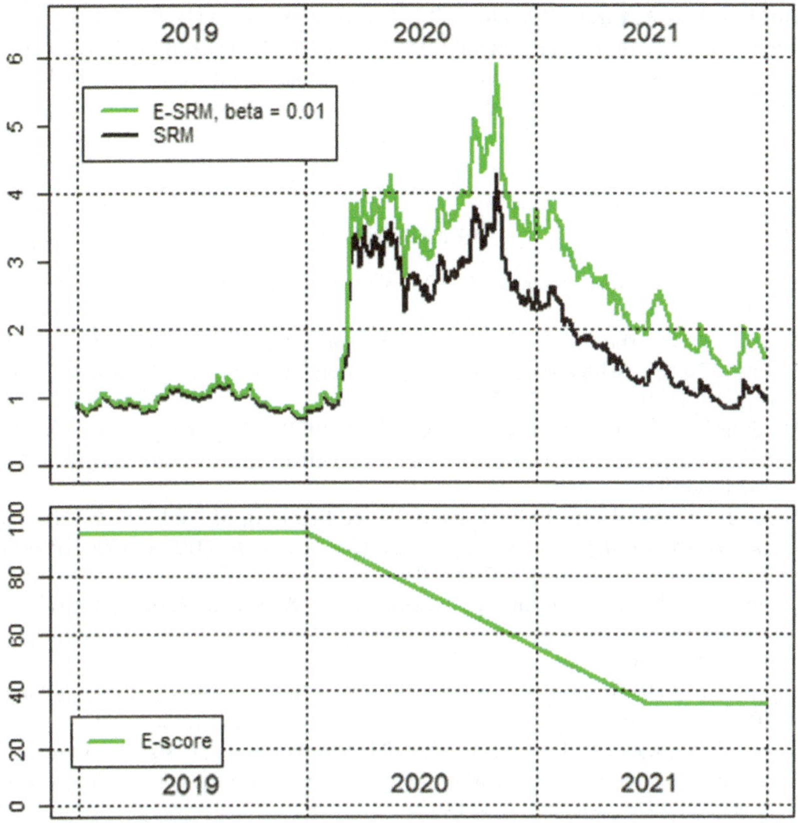

Fig. 4.16 SRM and E-SRM of the OSIIs most affected by the COVID-19 pandemic 2019–2021

With the increasing ESG data availability and improving quality of reporting of this data related to the upcoming IFRS reform, the utility of our solution will rise significantly. Finally, it should be noted that using the ESG data is not only cost-efficient but also the sole feasible solution for the frontier and emerging markets, where other data are very limited. Thanks to the global popularity of ESG scoring, our approach is a solution for measuring environmental risk exposure that is readily applicable to systemic risk analysis in both—developed and developing—markets.

Funding and Acknowledgments The National Science Centre funded this research project under Agreement UMO-2018/29/N/HS4/02783. We thank the participants of the 15th International Risk Management Conference and 21st CREDIT Conference for the insightful feedback that helped improve the paper significantly.

References

Adrian, T., & Brunnermeier, M. K. (2016). CoVaR. *American Economic Review, 106*, 1705–1741.

Aevoae, G.-M., Andries, A. M., Ongena, S., Sprincean, N., & ESG and Systemic Risk. (2022). Swiss Finance Institute Research Paper No. 22–25. https://doi.org/10.2139/ssrn.4058477

Albuquerque, R., Koskinen, Y., & Zhang, C. (2019). Corporate social responsibility and firm risk: Theory and empirical evidence. *Management Science, 65*, 4451–4469.

Alessi, L., Di Girolamo, F. E., Pagano, A., & Petracco, M. (2022). *Accounting for climate transition risk in banks' capital requirements* (JRC Working Papers in Economics and Finance, 2022/8).

Anginer, D., Demirguc-Kunt, A., Huizinga, H., & Ma, K. (2018). Corporate governance of banks and financial stability. *Journal of Financial Economics, 130*, 327–346.

Anginer, D., Demirguc-Kunt, A., & Zhu, M. (2014). How does competition affect bank systemic risk? *Journal of Financial Intermediation, 23*, 1–26.

Bahadori, N., Kaymak, T., & Seraj, M. (2021). Environmental, social, and governance factors in emerging markets: The impact on firm performance. *Business Strategy & Development, 4*, 411–422.

Bank for International Settlements (BIS). (2021, April). *Climate-related risk drivers and their transmission channels*. Basel Committee on Banking Supervision. https://www.bis.org/bcbs/publ/d517.pdf. Accessed 6 October 2022.

Bank of England (BoE). (2018, November). *Climate change: What are the risks to financial stability?* Knowledge Bank. https://www.bankofengland.co.uk/knowledgebank/climate-change-what-are-the-risks-to-financial-stability. Accessed 6 October 2022.

Bank of England (BoE). (2021, December). *Financial stability report*. https://www.bankofengland.co.uk/-/media/boe/files/financial-stability-report/2021/december-2021.pdf?la=en&hash=62FF3E7484FF0FD1AD650FE41A77D32B3750F8CF. Accessed 6 October 2022.

Bătae, O. M., Voicu, D., & Feleagă, L. (2020). Environmental, social, governance (ESG), and financial performance of European Banks. *Journal of*

Accounting and Management Information Systems, 19(3). https://doi.org/10.24818/jamis.2020.03003

Battiston, S. (2019). The importance of being forward-looking: Managing financial stability in the face of climate risk (Banque de France Financial Stability Review No. 23—June 2019—Greening the Financial System: The New Frontier). Banque de France, Paris: Financial. https://www.banque-france.fr/sites/default/files/media/2019/08/27/financial_stability_review_23.pdf. Accessed 6 October 2022.

Battiston, S., Dafermos, Y., & Monasterolo, I. (2021). Climate risks and financial stability. *Journal of Financial Stability, 54*, 100867.

Berg, F., Koelbel, J., & Rigobon, R. (2022). Aggregate confusion: The divergence of ESG ratings. *Review of Finance*. https://doi.org/10.1093/rof/rfac033

Billio, M., Costola, M., Hristova, I., Latino, C., & Pelizzon, L. (2021). Inside the ESG ratings: (Dis)agreement and performance. *Corporate Social Responsibility and Environmental Management, 28*(5), 1426–1445. https://doi.org/10.1002/csr.2177

Boffo, R., Marshall, C., & Patalano, R. (2020). *ESG investing: Environmental pillar scoring and reporting*. OECD. www.oecd.org/finance/esg-investing-environmental-pillar-scoring-and-reporting.pdf. Accessed 6 October 2022.

Boubaker, S., Cellier, A., Manita, R., & Saeed, A. (2020). Does corporate social responsibility reduce financial distress risk? *Economic Modelling, 91*, 835–851.

Bouslah, K., Kryzanowski, L., & M'Zali, B. (2018). Social performance and firm risk: Impact of the financial crisis. *Journal of Business Ethics, 149*, 643–669.

Breeden, S. (2022, April 28). *Macropru – fit for the future?* – speech at Lancaster University. https://www.bankofengland.co.uk/speech/2022/april/sarah-breeden-speech-at-lancaster-university-during-an-agency-visit-to-the-north-west. Accessed 6 October 2022.

Brownlees, C., & Engle, R. (2017). SRISK: A conditional capital shortfall measure of systemic risk. *Review of Financial Studies, 30*(1), 48–79.

Brunetti, C., Dennis, B., Gates, D., Hancock, D., Ignell, D., Kiser, E., Kotta, G., Kovner, A., Rosen, R., Tabor, N. (2021, March). *Climate change and financial stability, FEDS notes*. Board of Governors of the Federal Reserve System. https://www.federalreserve.gov/econres/notes/feds-notes/climate-change-and-financial-stability-20210319.htm. Accessed 6 October 2022.

Candelon, B., Hasse, J. B., & Lajaunie, Q. (2021). ESG-washing in the mutual funds industry? From information asymmetry to regulation. *Risks, 9*(11), 199. https://doi.org/10.3390/risks9110199

Cerqueti, R., Ciciretti, R., Dalò, A., & Nicolosi, M. (2021). ESG investing: A chance to reduce systemic risk. *Journal of Financial Stability, 54*, 100887.

CFA. (2022). *Systemic risk and management in finance*. Retrieved 23 September 2022. https://www.cfainstitute.org/en/advocacy/issues/systemic-risk

Chenet, H., Ryan-Collins, J., & van Lerven, F. (2021). Finance, climate-change and radical uncertainty: Towards a precautionary approach to financial policy. *Ecological Economics, 183*(May), 106957. https://doi.org/10.1016/j.ecolecon.2021.106957

Chiaramonte, L., Dreassi, A., Girardone, C., & Piserà, S. (2021). Do ESG strategies enhance bank stability during financial turmoil? Evidence from Europe. *European Journal of Finance*. https://doi.org/10.1080/1351847X.2021.1964556

Cormier, D., Ledoux, M., & Magnan, M. (2011). The informational contribution of social and environmental disclosures for investors. *Management Decision, 49*, 1276–1304.

Cornett, M. M., Erhemjamts, O., & Tehranian, H. (2016). Greed or good deeds: An examination of the relation between corporate social responsibility and the financial performance of U.S. commercial banks around the financial crisis. *Journal of Banking & Finance, 70*, 137–159.

Delis, M. D., de Greiff, K., Iosifidi, M., & Ongena, S. (2021). *Being stranded with fossil fuel reserves? Climate policy risk and the pricing of bank loans* (Swiss Finance Institute Research Paper Series No. 18–10).

Dziwok, E., & Karaś, M. (2021). Systemic illiquidity noise-based measure—A solution for systemic liquidity monitoring in frontier and emerging markets. *Risks, 7*, 124. https://doi.org/10.3390/risks9070124

Dziwok, E., Karaś, M., Stachura, M., & Szczepaniak, W. (2022). *E-SRISK – A method to quantify the environmental factor in systemic risk analysis*. 21st international conference CREDIT 2022 long run risks, Venice, 22–23 September. https://doi.org/10.2139/ssrn.4156777

Engle, R. F. (2022). *Long run risk and climate hedge portfolios – Invited talk*. 21st international conference CREDIT 2022 long run risks, Venice, 22–23 September.

Eratalay, M. H., & Cortés Ángel, A. P. (2022). The impact of ESG ratings on the systemic risk of european blue-chip firms. *Journal of Risk and Financial Management, 15*(4), 153. https://doi.org/10.3390/jrfm15040153

European Banking Authority (EBA). (2021). *Report on management and supervision of ESG risks for credit institutions and investment firms*. https://www.eba.europa.eu/sites/default/documents/files/document_library/Publications/Reports/2021/1015656/EBA%20Report%20on%20ESG%20risks%20management%20and%20supervision.pdf. Accessed 6 October 2022.

European Banking Authority (EBA). (2022a). *EBA publishes binding standards on Pillar 3 disclosures on ESG risks*. https://www.eba.europa.eu/eba-publishes-binding-standards-pillar-3-disclosures-esg-risks. Accessed 6 October 2022a.

European Banking Authority (EBA). (2022b). *O-SIIs, other systemically important institutions*. https://www.eba.europa.eu/risk-analysis-and-data/other-systemically-important-institutions-o-siis-. Accessed 6 October 2022b.

European Central Bank (ECB). (2009, December). *Financial stability review*. https://www.ecb.europa.eu/pub/pdf/fsr/art/ecb.fsrart200912_02.en.pdf. Accessed 6 October 2022.

European Central Bank (ECB). (2022). *ECB banking supervision launches 2022 climate risk stress test*. https://www.bankingsupervision.europa.eu/press/pr/date/2022/html/ssm.pr220127~bd20df4d3a.en.html?utm_source=ecb_twitter&utm_campaign=220127_pr_climate_stress_test. Accessed 6 October 2022.

European Central Bank (ECB). (2021a, November). *Financial stability review*. https://www.ecb.europa.eu/pub/pdf/fsr/ecb.fsr2021a11~8b0aebc817.en.pdf. Accessed 6 October 2022.

European Central Bank (ECB). (2021b, May). *Financial stability report*. https://www.ecb.europa.eu/pub/financial-stability/fsr/html/ecb.fsr2021b05~757f727fe4.en.html. Accessed 6 October 2022.

European Securities and Markets Authority (ESMA). (2022). *Sustainable finance roadmap 2022–2024*. https://www.esma.europa.eu/policy-activities/sustainable-finance/sustainable-finance-roadmap-2022-2024. Accessed 6 October 2022.

European Systemic Risk Board (ESRB). (2021). *Climate-related risk and financial stability*. https://www.esrb.europa.eu/pub/pdf/reports/esrb.climateriskfinancialstability202107~79c10eba1a.en.pdf?71a273dc36a85ef05c8bed530466f900. Accessed 6 October 2022.

Financial Stability Board (FSB). (2022, April). *Supervisory and regulatory approaches to climate-related risks interim report*. https://www.fsb.org/2022/04/supervisory-and-regulatory-approaches-to-climate-related-risks-interim-report/. Accessed 6 October 2022.

Financial Stability Oversight Council (FSOC). (2021). *Report on climate-related financial risk 2021*. https://home.treasury.gov/system/files/261/FSOC-Climate-Report.pdf. Accessed 6 October 2022.

Finger, M., Gavious, I., & Manos, R. (2018). Environmental risk management and financial performance in the banking industry: A cross-country comparison. *Journal of International Financial Markets, Institutions and Money, 52*, 240–261.

Gangi, F., Meles, A., D'Angelo, E., & Daniele, L. M. (2019). Sustainable development and corporate governance in the financial system: Are environmentally friendly banks less risky? *Corporate Social Responsibility and Environmental Management, 26*, 529–547.

Gillan, S. L., Koch, A., & Starks, L. T. (2021). Firms and social responsibility: A review of ESG and CSR research in corporate finance. *Journal of Corporate Finance, 66*, 101889.

International Financial Reporting Standards Foundation, IFRS. (2022). *ISSB receivers global response on proposed sustainability disclosure standards*. https://www.ifrs.org/news-and-events/news/2022/08/issb-receives-global-response-on-proposed-sustainability-disclosure-standards/. Accessed 5 October 2022.

International Monetary Fund. (2022). *Global financial stability report*. https://www.imf.org/en/Publications/GFSR/Issues/2022/04/19/global-financial-stability-report-april-2022. Accessed 6 October 2022.

Jajuga, K., Karaś, M., Kuziak, K., & Szczepaniak, W. (2017). Ryzyko systemu finansowego [Risk of the financial system]. In: *Materiały i Studia* (p. 329). National Bank of Poland. https://www.nbp.pl/publikacje/materialy_i_studia/ms329.pdf. Accessed 6 October 2022.

Jung, H., Engle, R. F., & Berner, R. (2021). Climate stress testing (FRB of New York Staff Report No. 977). Rev. June 2022. https://doi.org/10.2139/ssrn.3931516

Kahneman, D. (2013). *Thinking, fast and slow*. Macmillan.

Karaś, M., & Szczepaniak, W. (2020). Fragility or contagion? Properties of systemic risk in the selected countries of Central and East-Central Europe. In: K. Jajuga, H. Locarek-Junge, L. Orlowski & K. Staehr (Eds.). *Contemporary trends and challenges in finance. Springer proceedings in business and economics*. Springer. https://doi.org/10.1007/978-3-030-43078-8_19

Karaś, M., & Szczepaniak, W. (2021a). Systemic risk in selected countries of Western and Central Europe. In: K. Jajuga, H. Locarek-Junge, L. T. Orlowski, K. Staehr (Eds.). *Contemporary trends and challenges in finance. Springer proceedings in business and economics*. Springer. https://doi.org/10.1007/978-3-030-73667-5_10

Karaś, M., & Szczepaniak, W. (2021b). Three ways to improve the systemic risk analysis of the Central and Eastern European region using SRISK and CoVaR. *Journal of Credit Risk, 17*(2). https://ssrn.com/abstract=3908027

Kim, S., Lee, G., & Kang, H.-G. (2021). Risk management and corporate social responsibility. *Strategic Management Journal, 42*, 202–230.

KPMG International Limited. (2022). *IFRS announce new ESG reporting standard*. https://home.kpmg/dp/en/home/insights/2021/10/ifrs-announce-new-esg-reporting-standard.html. Accessed 5 October 2022.

Lamperti, F., Bosetti, V., Roventini, A., & Tavoni, M. (2019). The public costs of climate-induced financial instability. *Nature Climate Change, 9*(11), 829–833. https://doi.org/10.1038/s41558-019-0607-5

Lamperti, F., Bosetti, V., Roventini, A., Tavoni, M., & Treibich, T. (2021). Three green financial policies to address climate risks. *Journal of Financial Stability, 54*, 100875. https://doi.org/10.1016/j.jfs.2021.100875

Liu, W., Shao, X., De Sisto, M., & Li, W. H. (2021). A new approach for addressing endogeneity issues in the relationship between corporate social responsibility and corporate financial performance. *Finance Research Letters, 39*, 101623.

Mehta, A. (2022). *Symbols vs. Solutions: Quant as a driver for value creation.* Quant insights conference: ESG & climate risk in finance, CQF Institute, online, 29 September. https://hopin.com/events/quant-insights-conference-esg-climate-risk-in-quant-finance. Accessed 29 September 2022.

Murè, P., Spallone, M., Mango, F., Marzioni, S., & Bittucci, L. (2021). ESG and reputation: The case of sanctioned Italian banks. *Corporate Social Responsibility and Environmental Management, 28*(1), 265–277.

Neitzert, F., & Petras, M. (2021). Corporate social responsibility and bank risk. *Journal of Business Economics*. https://doi.org/10.1007/s11573-021-01069-2

Network for Greening the Financial System, NGFS. (2020, June). *The macroeconomic and financial stability impacts of climate change: Research priorities, technical report*. Bank of France. www.ngfs.net/sites/default/files/medias/documents/ngfs_research_priorities_final.pdf. Accessed 6 October 2022.

Nieto, M. (2017, October). *Banks and environmental sustainability: Some financial stability reflections*. International Research Centre on Cooperative Finance. https://doi.org/10.2139/ssrn.3082107

Reisinger, A. Howden, M., Vera, C., Garschagen, M., Hurlbert, M., Kreibiehl, S., Mach, K. J., Mintenbeck, K., O'Neill, B., Pathak, M., Pedace, R., Pörtner, H. S., Poloczanska, E., Corradi, M. R., Sillmann, J., van Aalst, M., Viner, D., Jones, R., Ruane, A. C., Ranasinghe, R. (2020). *The concept of risk in the IPCC sixth assessment report: A summary of cross-working group discussions*. Intergovernmental Panel on Climate Change. https://www.ipcc.ch/site/assets/uploads/2021/02/Risk-guidance-FINAL_15Feb2021.pdf. Accessed 6 October 2022.

Renneboog, L., Ter Horst, J., & Zhang, C. (2011). Is ethical money financially smart? Nonfinancial attributes and money flows of socially responsible investment funds. *Journal of Financial Intermediation, 20*, 562–588.

Sassen, R., Hinze, A.-K., & Hardeck, I. (2016). Impact of ESG factors on firm risk in Europe. *Journal of Business Economics, 86*, 867–904.

Scatigna M., Xia D., Zabai A., & Zulaica O. (2021, December). Achievements and challenges in ESG markets. *BIS Quarterly Review*. Bank for International Settlements.

Sohag, K., Hammoudeh, S., Elsayed, A., Mariev, O., & Safonova, Y. (2022). Do geopolitical events transmit opportunity or threat to green markets? Decomposed measures of geopolitical risks. *Energy Economics*, 106068. https://doi.org/10.1016/j.eneco.2022.106068

Tol, R. S. J. (2019). A social cost of carbon for (almost) every country. *Energy Economics, 83*, 555–566. https://doi.org/10.1016/j.eneco.2019.07.006

Toma, P., & Stefanelli, V. (2022). What are the banks doing in managing climate risk? Empirical evidence from a position map. *Ecological Economics, 200*(October), 107530. https://doi.org/10.1016/j.ecolecon.2022.107530

Wong, J. B., & Zhang, Q. (2022). Stock market reactions to adverse ESG disclosure via media channels. *The British Accounting Review, 54*, 101045.

Zhang, X., Zhang, S., & Lu, L. (2022). The banking instability and climate change: Evidence from China. *Energy Economics, 106*, 105787. https://doi.org/10.1016/j.eneco.2021.105787

CHAPTER 5

Corruption Disclosure in Banking: Insights from the Literature

Pablo de Andrés, Salvatore Polizzi, Enzo Scannella, and Nuria Suárez

5.1 Introduction

Transparency and disclosure are pivotal to attenuate the adverse effects of information asymmetries that strongly influence the relationship between banks and their stakeholders (Levine, 1997) and to ensure an adequate

P. de Andrés
Finance and Marketing Department, Universidad Autónoma de Madrid and ECGI, Madrid, Spain
e-mail: p.andres@uam.es

S. Polizzi (✉) · E. Scannella
Department of Economics, Business and Statistics, University of Palermo, Palermo, Italy
e-mail: salvatore.polizzi@unipa.it

E. Scannella
e-mail: enzo.scannella@unipa.it

P. Wachtel et al. (eds.), *Creating Value and Improving Financial Performance*, Palgrave Macmillan Studies in Banking and Financial Institutions, https://doi.org/10.1007/978-3-031-24876-4_5

and efficient functioning of both banking and financial markets. Among the most important benefits of transparency in the banking sector, it is noteworthy to mention its reduction effect on the cost of capital; its relationship with higher levels of stakeholders' and investors' trust (Botosan & Plumlee, 2002; Eng & Mak, 2003); the more effective bank lending activity (Zelenyuk et al., 2020); and the higher levels of financial stability (Nier, 2005).

Although the extant literature has remarked the importance of financial disclosure and of the disclosure of the most traditional types of banking risks (Birindelli & Ferretti, 2017; Frolov, 2006; Giner et al., 2020; Pérignon & Smith, 2010), also non-financial disclosure plays an important role. Over the last decades, stakeholders have demanded higher levels of corporate social responsibility by banks (García-Sánchez et al., 2018; Pérez & del Bosque, 2012), which are required to be more transparent on various aspects that are not related to financial disclosure, including their impact on the environment, on the society, and on a wide range of non-financial dimensions. These non-financial CSR dimensions include community involvement, human resources, environment (Farina et al., 2019; Kiliç, 2016; Schröder, 2021), and possible involvements in corruption scandals and anti-corruption mechanisms (de Andrés et al., 2022).

Some studies have examined corruption disclosure in non-financial firms (e.g., Blanc et al., 2017, 2019; Joseph et al., 2016), showing that while firms generally provide lower levels of corruption disclosure, this type of information can be used to restore their institutional legitimacy in the eyes of their stakeholders after the occurrence of a corruption event. On the other hand, with specific reference to the banking industry, previous research has largely ignored this topic, notwithstanding its theoretical and practical relevance. To the best of our knowledge, de Andrés et al. (2022) is the only attempt to examine corruption disclosure in the banking sector. De Andrés et al. (2022), based on the signaling theory

---

N. Suárez
Finance and Marketing Department, Universidad Autónoma de Madrid, Madrid, Spain
e-mail: nuria.suarez@uam.es

(Spence, 1973), argue that banks use corruption disclosure to send a signal to stakeholders and investors concerning the reliability of their anti-corruption mechanisms and highlight the fact that they have not been involved in corruption scandals. It is surprising that this is the only study focusing on this specific topic, considering the negative consequences that corruption events could have in a single bank and the *domino* effect that could be triggered to the whole banking system. Moreover, corruption scandals may jeopardize banks' reputation and, in a wider perspective, the stability of the financial system and its ability to adequately support the real economy.

In this chapter, we aim to review the strand of literature focusing on corruption disclosure and to describe the most important theoretical frameworks that can be adopted to examine this type of disclosure with specific reference to the banking industry. The analysis of the literature on corruption disclosure in non-financial firms represents an important step to pave the way for future research focusing on the banking industry. In addition, the identification of the most suitable theoretical frameworks that can be used to interpret the findings of empirical studies can support researchers in defining their research design to advance our knowledge in this field of research.

We contribute to the literature by bridging the gap between the literature focusing on corruption disclosure and the banking literature. In doing so, we pave the way for future cutting-edge research that studies the determinants, consequences, and theoretical justification of corruption disclosure in banking.

The remainder of this chapter is structured as follows: Sect. 5.2 provides a general background on the effects of corruption on the banking industry and more specifically on the lending business. Section 5.3 describes the theoretical frameworks that can be adopted to analyze corruption disclosure in the banking sector. Section 5.4 reviews the strand of literature focusing on corruption disclosure in financial and non-financial firms. Lastly, Sect. 5.5 concludes.

5.2 Corruption on the Banking Industry and the Lending Business

The debate on the effects of corruption on economic growth and that related to the channels of this broad relationship has captured scholars' and policymakers' attention for decades (Gründler & Potrafke, 2019; Mauro, 1995; Mo, 2001). Considering the widely acknowledged importance of the banking and financial sector as a fundamental determinant of economic growth, the literature has also investigated into the consequences of corruption on the banking industry and, in particular, on the lending business. On the one hand, some studies shed light on the negative consequences that corruption has on economic performance and on the well-functioning of the banking/lending channel. For instance, Park (2012) finds that corruption may cause a distortion of bank lending from good to bad projects, which results in a lower quality of private investments and consequently in lower rates of economic growth. In addition, corruption causes a deterioration of bank loan quality, making the financial system less stable and more exposed to financial crises. Similar results have also been obtained by Son et al. (2020) and Goel and Hasan (2011) who show that bank lending and, particularly, the levels of non-performing loans are positively related to the level of corruption. Among the scholars that have remarked the negative consequences of corruption in the banking industry, it is noteworthy to mention also Asteriou et al. (2021) who show that corruption has a negative influence on bank stability and profitability. On the other hand, although it is a minoritarian strand of the literature, other authors advocate that corruption can have neutral or even positive effects on economic development by means of the banking channel. For instance, Song et al. (2021) suggest that developing countries can temporarily indulge corruption while implementing policies to increase financial development to promote economic growth. Mongid and Tahir (2011) even provide evidence of a positive relationship between corruption and bank profitability in developing countries, shedding light on the possible positive consequences of corruption in the banking industry.

The relationship between corruption and bank lending is influenced by several factors. On the one hand, some authors find that country-level characteristics play a prominent role. For instance, Zheng et al. (2013) show that the relationship between corruption and the lending business is particularly strong in collectivist countries and relatively weaker in individualist countries. In addition, the country-level institutional framework has emerged as a particularly relevant aspect, including the supervisory style of bank regulatory authorities. In particular, Beck et al. (2006) show that when bank supervisors rely more on market discipline mechanisms to monitor bank activity, then the degree to which corrupted bank officials negatively affect firm ability to access external finance is reduced. In contrast, more traditional supervisory styles are associated with more severe problems related to corruption in the lending business. On a more general level, Barry et al. (2016) find that a stronger regulatory environment reduces bank lending corruption. Akins et al. (2017) find that also country-level accounting policies in terms of timely loan loss recognition reduce lending corruption as it significantly increases the chance that a problematic loan can be identified earlier. On the other hand, also bank level characteristics influence the relationship between lending and corruption. In this vein, Weill (2011a) shows that when banks are particularly risk averse, corruption can even have a beneficial effect on bank lending, supporting the idea that corruption can play a role in alleviating distortions determined by inefficient and ineffective institutions (the so-called greasing the wheels hypothesis (Chen et al., 2013). Weill (2011b) demonstrates that corruption has a detrimental effect for those banks that lend more to firms and households, while lending to government is not affected. Bermpei et al. (2021) find that for those banks that engage more in relationship lending, the negative effect of corruption on lending is weaker.

In sum, given the mixed and to some extent contrasting results of the literature, we remark the importance of further research in this field of study, aiming to shed light on the actual consequences of corruption on the banking and lending business.

5.3 The Importance of Corruption Disclosure

There are numerous reasons why corruption disclosure is particularly important. For instance, companies may be willing to provide this type of disclosure as a means to respond to stakeholders' social and political pressure or to project an image of action and awareness of corruption problems. This would be particularly relevant especially after the occurrence of corruption scandals. On the one hand, those companies (banks) that were involved in corruption problems may be willing to restore their institutional legitimacy by means of corruption disclosure, while on the other hand also those entities that were not involved in these types of issues may be willing to send a signal to investors and shareholders of the lack of corruption problems. Transparency with reference to corruption problems is particularly important in the banking industry because corruption events can harm bank reputation and image, with potential negative consequences in terms of profitability in the long run (Altunbaş et al., 2018). If investors lose their confidence in the integrity of the banking and financial system, the effectiveness of financial intermediation could be negatively affected. Hence, it is important that current and potential investors and stakeholders are sufficiently informed about any type of involvement in corruption problems of the bank and its anti-corruption mechanisms in place to deal with these situations.

The strand of literature focusing on non-financial firms offers important insights to study more in-depth corruption disclosure and represents an important point of reference for future research in this field. Among these studies, Blanc et al. (2017) explore the relationships between anti-corruption disclosure and media exposure. Specifically, these authors examine the Transparency International's[1] Rating of the Anti-Corruption Disclosure of 105 large multinational companies at international level and the Reporters Without Borders' rankings of state-level press freedom.[2] The main findings of their study are that media exposure is positively associated with the aforementioned scores of anti-corruption disclosure, and that this positive association is weaker if the levels of press freedom are high. Blanc et al. (2019) provide a detailed analysis of the anti-corruption disclosure at Siemens AG both before and after the spread of the news of a corruption event occurred in 2006. These authors analyze this case study under the lenses of the stakeholder (Freeman, 1984) and legitimacy theories (Suchman, 1995). They show that the occurrence of the scandal is positively related with increased levels of corruption disclosure. This result would be consistent with an attempt to restore the institutional legitimacy of the company after the corruption event. From a country-level perspective, Gago-Rodríguez et al. (2020) provide evidence that firms operating within more corrupt legal environments, facing more competition, and bearing a higher risk of being identified are less likely to deny

[1] https://www.transparency.org/en.

[2] http://en.rsf.org/press-freedom-index-2011-2012,1043.html.

their involvement in bribery. Sari et al. (2021) develop an empirical analysis on anti-corruption disclosure by using a disclosure index constructed by drawing upon the global reporting initiative. Their main finding is that, notwithstanding the efforts of several international initiatives to increase the level and homogeneity of corruption disclosure across countries, there are still significant differences at country level and the amount of disclosure is generally low. This result is likely due to the fact that corruption disclosure is considered a highly sensitive information, and therefore firm managers are very careful in increasing transparency in such a delicate and confidential matter.

To the best of our knowledge, the contribution proposed by de Andrés et al. (2022) is the only one focusing on corruption disclosure in the banking sector. These authors carry out an empirical investigation of corruption disclosure by analyzing a sample of 88 banks headquartered in the so-called GIPSI countries that include Greece, Ireland, Portugal, Spain, and Italy. Their research design is based on the analysis of 22 banks that have been involved in corruption events ("corrupted banks") and 66 banks uninvolved in such a type of problem. De Andrés et al. (2022) employ an automated content analysis methodology (Krippendorff, 2004; Weber, 1990) and analyze the annual financial reports of these banks by using a tailored corruption disclosure dictionary validated by a panel of experts. Their main finding is that "uncorrupted banks" are more transparent compared to "corrupted banks", even after the occurrence of corruption scandals. This result is in line with the signaling theory (Spence, 1973), in that "uncorrupted banks" disclose more information related to corruption, in an attempt to send a signal to market participants and stakeholders regarding the effectiveness of their internal anti-corruption mechanisms and the fact that they have not been involved in any corruption event.

Broadly speaking, it is important to highlight that corruption events are much more important for banking institutions compared to non-financial firms. Reputation is one of the most important assets for any bank (Bahoo, 2020; Johnson, 1997) and it is fundamental for their survival and growth. Corruption scandals might negatively affect bank reputation and trigger a *domino* effect that could result in negative consequences for the whole financial system. For these reasons, banks should be transparent on this regard and provide adequate levels of corruption disclosure.

5.4 Theoretical Frameworks for Corruption Disclosure

In order to investigate the disclosure behavior related to corruption, it is important to rely on adequate theoretical frameworks that could explain firm's and, more precisely, bank's corruption disclosure practices. Among the various theoretical frameworks that are adopted in the disclosure literature, three theories are particularly useful to explain corruption disclosure, namely: (i) stakeholder theory; (ii) legitimacy theory; and (iii) signaling theory.

The stakeholder theory (Freeman, 1984) posits that stakeholders have a significant influence on managerial decisions because the most important purpose of any firm is to satisfy stakeholders' needs. In addition, considering that firms must use the resources necessary to produce their goods or services, and that these resources are directly or indirectly controlled by stakeholders, it is clear that the more resources are controlled by these stakeholders (i.e., the higher their power), the more they influence firms' behavior. When it comes to disclosure practices, this theory posits that firms use disclosure in order to interact and communicate with their most influential stakeholders, and to get access to the aforementioned resources. This theoretical framework is strictly intertwined with legitimacy theory (Suchman, 1995) because disclosure can be considered a tool to achieve high levels of institutional legitimacy among stakeholders. With specific reference to corruption disclosure, given that the occurrence of corruption events and the anti-corruption mechanisms implemented by firms are particularly important, firms can use this type of disclosure to interact with their stakeholders for two main reasons. The first reason is to restore institutional legitimacy after the occurrence of a corruption scandal. Hence, firms can provide higher levels of corruption disclosure to restore stakeholders' confidence in the firm and inform them about the fact that they are putting effort to solve these corruption problems, and that they are committed to avoid other corruption scandals in future. The second reason is that after the occurrence of corruption scandals, firms that were not involved in these types of problem are willing to inform their stakeholders about their lack of involvement in corruption scandals and about the effectiveness of their anti-corruption policies. In sum, according to this theory, both firms that have been involved in corruption events and those that were not involved may be willing to provide higher levels of corruption disclosure after the occurrence of a corruption scandal.

The theoretical perspective proposed by the legitimacy theory allows disentangling better these two possible effects. The legitimacy theory, originally proposed by Suchman (1995), helps to understand why firms demonstrate their adherence to the system of values of the society and how they meet the expectations of the society. More specifically, firms are bound into a social contract (which can be expressed or implied) with the society, whereby their growth and survival depend on the delivery of socially desirable goods and/or services. If firms do not respect this contract, they should put a remedy to this situation, and they can do it by providing higher levels of disclosure to explain the activities they carry out and to show their adherence to the social contract. When it comes to corruption disclosure, the legitimacy theory posits that if firms are involved in corruption scandals, they should provide higher levels of disclosure to restore their legitimacy. In contrast, firms that were not involved in corruption problems should not change their disclosure behavior, as their institutional legitimacy has not been affected. Hence, the legitimacy theory can be considered (at least partially) in line with stakeholder theory, when it comes to corruption disclosure.

A completely different perspective is offered by the signaling theory proposed by Spence (1973). Although this theory is generally used to explain firms' behavior with reference to the disclosure on financial performance (Dicuonzo, 2018), it can be also used for non-financial disclosure, in general, and for corruption disclosure, in particular. According to this theory, high-performing firms are willing to provide more information to the market about their levels of financial performance, and they do it by means of disclosure. A similar argument holds with reference to the disclosure on corruption. Firms that have reliable and effective anti-corruption mechanisms should be willing to send a signal to investors to show the effectiveness of their anti-corruption policies. In addition, after the occurrence of a corruption scandal, firms that were not involved in this type of events are incentivized to disclose information about their lack of involvement in corruption problems. Therefore, they are supposed to provide higher levels of corruption disclosure than firms that experienced corruption issues. It is clear that this theory is in stark contrast with the legitimacy theory. The understanding on which of the theories is more effective to explain corruption-related disclosure is an empirical question that cannot be answered without an adequate empirical analysis.

Although the empirical analysis of corruption disclosure would be a suitable strategy to detect the most important theories to explain bank disclosure practices, there is an important challenge to be tackled. To the best of our knowledge, there is no methodological approach that has been specifically designed to analyze corruption disclosure in the banking industry. In order to solve this problem, we suggest developing qualitative judgment-based metrics or dictionaries of terms related to corruption that could be used to analyze this type of disclosure. The extant literature has recently shown that both approaches are perfectly suitable to investigate into bank disclosure practices (Altunbaş et al., 2022; Scannella & Polizzi, 2021).

5.5 Conclusions

This chapter proposed a review of the most important contributions on corruption-related disclosure and of the theoretical frameworks that can be employed in this field of study, aiming to bridge the gap with the banking literature, which has ignored this topic so far.

Among the various dimensions of CSR disclosure (i.e., environmental disclosure, disclosure on human resources, and community involvement), corruption disclosure deserves particular attention, although it has not been analyzed in depth in the extant literature. While there are some studies that analyze this topic by focusing on non-financial firms (e.g., Blanc et al., 2017, 2019; Joseph et al., 2016), to the best of our knowledge, one single paper focuses on the banking industry (de Andrés et al., 2022).

Corruption disclosure is particularly important not only for firms in general, but also with specific reference to the banking sector. Broadly speaking, firms can use this type of disclosure to respond to stakeholders' social and political pressure or to project an image of action and awareness of corruption problems, and it is particularly relevant especially after the occurrence of corruption scandals that sometimes are experienced by financial and non-financial firms (Bahoo, 2020; Bahoo et al., 2020; Blanc et al., 2019). Transparency on corruption problems is particularly important in the banking sector, as corruption events can harm bank image and reputation, with negative consequences also in terms of profitability (Altunbaş et al., 2018). Hence, banks can use this disclosure to enhance and restore their reputation after these scandals.

Given the importance of corruption disclosure, it is particularly important to fill this gap in the literature and offer theoretical and empirical contributions. In order to do so, scholars can rely on three main theoretical frameworks, namely: stakeholder theory (Freeman, 1984); legitimacy theory (Suchman, 1995), and signaling theory (Spence, 1973). Given the stark contrast between the legitimacy theory and the signaling theory, especially when it comes to the analysis of the effects of corruption events, detailed empirical analyses of the banking industry are necessary to understand which of the two theories play the most prominent role in explaining corruption disclosure practices. Thus, apart from the practical importance of the analysis of corruption disclosure, there are also relevant theoretical aspects that require further investigations in the literature.

In conclusion, in this chapter we highlighted the importance of corruption disclosure and paved the way for future research in this field of studies. This research area is promising from both a theoretical and a practical viewpoint. On the one hand, it is important to understand whether legitimacy theory, stakeholder theory, or signaling theory can explain the relationship between corruption scandals and corruption disclosure in banking. On the other hand, from a practical perspective, this field of research is useful to understand the actual disclosure strategies and practices banks adopt when it comes to corruption events and anti-corruption mechanisms and processes, as well as the real effects associated with the different disclosure policies on corruption issues. We hope that this piece of research will give momentum to the literature on corruption disclosure in the banking sector.

Acknowledgements We acknowledge participants at the EUMODFRAUD Final Conference (Porto), at the XXX ACEDE Conference (Cartagena), at the 2021 Wolpertinger Conference (Cracow), and at the 14th IRMC Conference (Cagliari). This research benefited from the Professorship Excellence Program in accordance with the multi-year agreement signed by the Government of Madrid and the Autonoma University of Madrid (Line #3). We acknowledge financial support from the EU PROJECT H2020 and Hercules III Program (HERCULE-2019-LT-AG). P. de Andrés, E. Scannella and N. Suárez acknowledge financial support from the Spanish Ministry of Economy and Competitiveness (Project PID2020-118064GB-I00). S. Polizzi and N. Suárez acknowledge financial support from the Comunidad de Madrid Research Project for Young Researchers (SI3-PJI-2021-00276).

References

Akins, B., Dou, Y., & Ng, J. (2017). Corruption in bank lending: The role of timely loan loss recognition. *Journal of Accounting and Economics, 63*(2–3), 454–478.

Altunbaş, Y., Polizzi, S., Scannella, E., & Thornton, J. (2022). European Banking Union and bank risk disclosure: The effects of the Single Supervisory Mechanism. *Review of Quantitative Finance and Accounting, 58*(2), 649–683.

Altunbaş, Y., Thornton, J., & Uymaz, Y. (2018). CEO tenure and corporate misconduct: Evidence from US banks. *Finance Research Letters, 26*, 1–8.

Asteriou, D., Pilbeam, K., & Tomuleasa, I. (2021). The impact of corruption, economic freedom, regulation and transparency on bank profitability and bank stability: Evidence from the Eurozone area. *Journal of Economic Behavior and Organization, 184*, 150–177.

Barry, T. A., Lepetit, L., & Strobel, F. (2016). Bank ownership structure, lending corruption and the regulatory environment. *Journal of Comparative Economics, 44*(3), 732–751.

Bahoo, S. (2020). Corruption in banks: A bibliometric review and agenda. *Finance Research Letters, 35*, 101499.

Bahoo, S., Alon, I., & Paltrinieri, A. (2020). Corruption in international business: A review and research agenda. *International Business Review, 29*(4), 101660.

Beck, T., Demirgüç-Kunt, A., & Levine, R. (2006). Bank supervision and corruption in lending. *Journal of Monetary Economics, 53*(8), 2131–2163.

Bermpei, T., Kalyvas, A. N., & Leonida, L. (2021). Local public corruption and bank lending activity in the United States. *Journal of Business Ethics, 171*(1), 73–98.

Birindelli, G., & Ferretti P. (2017). Operational risk management in banks: Regulatory, organisational and strategic issues. In P. Molyneux (Ed.), *Palgrave Macmillan studies in banking and financial institutions*. Palgrave Macmillan.

Blanc, R., Islam, M. A., Patten, D. M., & Branco, M. C. (2017). Corporate anti-corruption disclosure: An examination of the impact of media exposure and country-level press freedom. *Accounting, Auditing and Accountability Journal, 30*(8), 1746–1770.

Blanc, R., Cho, C. H., Sopt, J., & Branco, M. C. (2019). Disclosure responses to a corruption scandal: The case of Siemens AG. *Journal of Business Ethics, 156*(2), 545–561.

Botosan, C. A., & Plumlee, M. A. (2002). A re-examination of disclosure level and the expected cost of equity capital. *Journal of Accounting Research, 40*(1), 21–40.

Chen, Y., Liu, M., & Su, J. (2013). Greasing the wheels of bank lending: Evidence from private firms in China. *Journal of Banking and Finance, 37*(7), 2533–2545.

de Andrés, P., Polizzi, S., Scannella, E., & Suárez, N. (2022, forthcoming). Corruption-related disclosure in the banking industry: Evidence from GIPSI countries. *The European Journal of Finance*. https://doi.org/10.1080/1351847X.2022.2153073

Dicuonzo, G. (2018). *La disclosure sui rischi finanziari tra dottrina, normativa e prassi: Evidenze empiriche dal contesto italiano*. Giappichelli.

Eng, L. L., & Mak, Y. T. (2003). Corporate governance and voluntary disclosure. *Journal of Accounting and Public Policy, 22*(4), 325–345.

Farina, V., Galloppo, G., & Previati, D. A. (2019). Central Banks' commitment to stakeholders: CSR in the eurosystem: 2006–2016. In E. Gualandri, V. Venturelli, & A. Sclip (Eds.), *Frontier topics in banking* (pp. 189–221). Palgrave Macmillan.

Freeman, R. E. (1984). *Strategic management: A stakeholder approach*. Cambridge University Press.

Frolov, M. (2006). Bank credit risk disclosure in Japan. *Journal of Banking Regulation, 7*(3–4), 221–242.

Gago-Rodríguez, S., Márquez-Illescas, G., & Núñez-Nickel, M. (2020). Denial of corruption: Voluntary disclosure of bribery information. *Journal of Business Ethics, 162*, 609–626.

García-Sánchez, I. M., Martínez-Ferrero, J., & García-Meca, E. (2018). Board of directors and CSR in banking: The moderating role of bank regulation and investor protection strength. *Australian Accounting Review, 28*(3), 428–445.

Giner, B., Allini, A., & Zampella, A. (2020). The value relevance of risk disclosure: An analysis of the banking sector. *Accounting in Europe, 17*(2), 129–157.

Goel, R. K., & Hasan, I. (2011). Economy-wide corruption and bad loans in banking: International evidence. *Applied Financial Economics, 21*(7), 455–461.

Gründler, K., & Potrafke, N. (2019). Corruption and economic growth: New empirical evidence. *European Journal of Political Economy, 60*, 101810.

Johnson, S. A. (1997). The effect of bank reputation on the value of bank loan agreements. *Journal of Accounting, Auditing and Finance, 12*(1), 83–100.

Joseph, C., Gunawan, J., Sawani, Y., Rahmat, M., Noyem, J. A., & Darus, F. (2016). A comparative study of anticorruption practice disclosure among Malaysian and Indonesian Corporate Social Responsibility (CSR) best practice companies. *Journal of Cleaner Production, 112*, 2896–2906.

Kiliç, M. (2016). Online corporate social responsibility (CSR) disclosure in the banking industry: Evidence from Turkey. *International Journal of Bank Marketing, 34*(4), 550–569.

Krippendorff, K. (2004). *Content analysis: An introduction to its methodology*. Sage.

Levine, R. (1997). Financial development and economic growth: Views and agenda. *Journal of Economic Literature, 35*(2), 688–726.

Mauro, P. (1995). Corruption and growth. *The Quarterly Journal of Economics, 110*(3), 681–712.

Mo, P. H. (2001). Corruption and economic growth. *Journal of Comparative Economics, 29*(1), 66–79.

Mongid, A., & Tahir, I. M. (2011). Impact of corruption on banking profitability in ASEAN countries: An empirical analysis. *Banks and Banking Systems, 6*(1), 41–48.

Nier, E. W. (2005). Bank stability and transparency. *Journal of Financial Stability, 1*(3), 342–354.

Park, J. (2012). Corruption, soundness of the banking sector, and economic growth: A cross-country study. *Journal of International Money and Finance, 31*(5), 907–929.

Pérez, A., & Del Bosque, I. R. (2012). The role of CSR in the corporate identity of banking service providers. *Journal of Business Ethics, 108*(2), 145–166.

Pérignon, C., & Smith, D. R. (2010). The level and quality of Value-at-Risk disclosure by commercial banks. *Journal of Banking and Finance, 34*(2), 362–377.

Sari, T. K., Cahaya, F. R., & Joseph, C. (2021). Coercive pressures and anti-corruption reporting: The case of ASEAN countries. *Journal of Business Ethics, 171*(3), 495–511.

Scannella, E., & Polizzi, S. (2021). How to measure bank credit risk disclosure? Testing a new methodological approach based on the content analysis framework. *Journal of Banking Regulation, 22*(1), 73–95.

Schröder, P. (2021). Corporate social responsibility (CSR) website disclosures: Empirical evidence from the German banking industry. *International Journal of Bank Marketing, 39*(5), 768–788.

Son, T. H., Liem, N. T., & Khuong, N. V. (2020). Corruption, nonperforming loans, and economic growth: International evidence. *Cogent Business and Management, 7*(1), 1735691.

Song, C. Q., Chang, C. P., & Gong, Q. (2021). Economic growth, corruption, and financial development: Global evidence. *Economic Modelling, 94*, 822–830.

Spence, M. (1973). Job market signaling. *The Quarterly Journal of Economics, 87*(3), 355–374.

Suchman, M. C. (1995). Managing legitimacy: Strategic and institutional approaches. *Academy of Management Review, 20*(3), 571–610.

Weber, R. P. (1990). *Basic content analysis*. Sage.

Weill, L. (2011a). How corruption affects bank lending in Russia. *Economic Systems, 35*(2), 230–243.
Weill, L. (2011b). Does corruption hamper bank lending? Macro and micro evidence. *Empirical Economics, 41*(1), 25–42.
Zelenyuk, N., Faff, R., & Pathan, S. (2020). The impact of voluntary capital adequacy disclosure on bank lending and liquidity creation. *Accounting and Finance, 60*(4), 3915–3935.
Zheng, X., Ghoul, S. E., Guedhami, O., & Kwok, C. C. (2013). Collectivism and corruption in bank lending. *Journal of International Business Studies, 44*(4), 363–390.

CHAPTER 6

Financial Competence and the Role of Non-cognitive Factors

Elisa Bocchialini, Beatrice Ronchini, and Francesca Torti

6.1 Introduction

Financial literacy has been recognized as a core life skill for young people in modern society (Lusardi, 2015; OECD, 2020). Mastery of financial concepts is increasingly seen as an essential precursor to financial well-being and active citizenship; furthermore, during these times plagued by

E. Bocchialini (✉)
Department of Economics and Management,
University of Florence, Florence, Italy
e-mail: elisa.bocchialini@unifi.it

E. Bocchialini · B. Ronchini
Department of Economics and Management, University of Parma, Parma, Italy
e-mail: beatrice.ronchini@unipr.it

F. Torti
European Commission, Joint Research Centre (JRC), Varese, Italy
e-mail: francesca.torti@ec.europa.eu

P. Wachtel et al. (eds.), *Creating Value and Improving Financial Performance*, Palgrave Macmillan Studies in Banking and Financial Institutions, https://doi.org/10.1007/978-3-031-24876-4_6

the COVID-19 pandemic investment in financial education seems also able to reduce the NEET status, build inclusiveness and financial resilience (Aina et al., 2021; Lyons et al., 2020). It follows that a truly thorough comprehension of the financial education process is quite crucial and thus represents a top priority worldwide as financial illiteracy remains widespread at a global level regardless of the proliferation of financial education programs. On this point, in Chapter 1, the authors show how, despite the proliferation of educational programs, the financial literacy of Italian adults deteriorated over the period under consideration.

Against the backdrop of the discussion above, academic efforts worldwide are still needed to understand "how people acquire and 'deploy' financial literacy" (Bongini et al., 2015). In this context, the study of the "attitudinal variables" received substantial academic attention in recent years, since they have been acknowledged as a fundamental component of financial literacy (OECD INFE, 2011) and thus recognized as an influential factor in financial learning (OECD, 2019). This notwithstanding, the available empirical evidence from this flourishing area of investigation is quite ambiguous yet, probably also due to the huge variety of definitions and measures that have been adopted so far across surveys in the absence of a shared conceptual framework (Nicolini, 2019; Remund, 2010).

Despite the variety of existing definitions, three main approaches seem to prevail in doctrine. The first, mainly proposed by the scholars who first addressed the issue of non-cognitive skills in the financial literacy field, tends to elude the definitional problem. Several academics within this stream of scholarship avoid giving ex-ante an explicit definition of financial attitude, which can be only inferred ex-post through the instruments used to assess it (Bocchialini et al., 2013). The second, supported by OECD/INFE (2015), looks at financial attitudes as one of the three components of financial literacy, the one which is basically "meant to capture attitudes towards precautionary saving" and to longer-term financial planning (D'Alessio et al., 2020). The third, followed by several scholars, focus on some people's traits—such as their vision of finance, their feelings towards finance, and their financial self-confidence—to understand whether these factors interfere with personal financial literacy (Danes & Haberman, 2007; Dobni & Racine, 2016; Van Der Cruijsen et al., 2021). Based on this approach, the above variables have typically been explored in isolation rather than in their interrelationship: they have not been included in a complete unifying framework and least of all under the banner of attitude.

In the face of this definitional challenge, no surprise that the role of attitude in the financial education process remains somewhat controversial. Does financial attitude influence financial knowledge? Does financial knowledge influence financial knowledge? Does financial knowledge influence financial attitude? Are they two independent constructs? How and to what extent could financial education initiatives that pay attention to learners' attitudinal profiles help financial learning? This issue is far from clear, probably also because most studies have so far mainly investigated the relationship between financial knowledge and financial behavior (Almenberg & Säve-Söderbergh, 2011; Van Rooij et al., 2012), rather than focusing on attitude, maybe also due to its evanescent nature (Riitsalu et al., 2019). Against this background, interesting new research opportunities open up. Just moving from this knowledge gap on the relationship between the two above variables, the present chapter seeks to evaluate if and to what extent attitude towards finance is important to predict financial knowledge or vice versa.

We define financial knowledge—intended as the cognitive component of financial literacy—as the basic understanding of those financial concepts that allow an individual to make responsible financial decisions (Huston, 2010). Inspired by Di Martino and Zan (2011), we also focus on the concept of "attitude towards finance", which we purposely adopt to mark the difference from the more popular construct of financial attitude (see, for instance, D'Alessio et al., 2020 and Talwar et al., 2021). We refer to "attitude towards finance" as the inclination of an individual to respond favorably or unfavorably to a particular financial stimulus, due to the durable mix of feelings (affective response) and opinions (cognitive response) held around financial matters. Unlike the OECD methodology, our definition does not merely capture the tendency of an individual "to look at financial issues in a long-term perspective" (D'Alessio et al., 2020), whereas it's still quite similar to others proposed in the financial literature (Garber & Koyama, 2016). In this study, we consider the attitude towards finance as a multifaced concept based on a combination of three dimensions: the view of finance, the emotional disposition towards finance, and the perceived competence in finance (Di Martino & Zan, 2011). Hence, having a favorable attitude towards finance essentially means that an individual holds an emotional system and a belief system (both towards finance and on self) that makes him or her open towards finance-related issues and willing to engage with them. In more detail, a positive attitude towards finance is a complex construct that conjointly implies that an individual: (1) has a favorable view (thoughts, opinions, and judgments) of finance. For example, the individual acknowledges the social and economic value of the financial sector and trusts financial institutions. He may also display a positive vision of finance as a field of study, for instance, believing in the usefulness, relevance, and worth of financial education in professional and personal life. (2) Enjoys finance learning experiences and finance-related activities and has no (or low) anxiety towards them. The individual may also hold aspirations towards finance and show interest in finance-related activities or in pursuing a finance-related career. (3) Has self-esteem in finance and is confident of succeeding in finance-related issues ("I'm able", "I can easily understand, learn and use finance", "I can succeed in finance-related issues").

Based on the above definition of attitude towards finance, this study surveys a sample of university students attending a public university located in the North of Italy in the attempt to measure their attitudes towards financial issues in conjunction with their level of financial knowledge. University students are generally considered "a particularly interesting group to study about financial capability issues" (Suyanto et al., 2021). Like other countries, in Italy they have been systematically investigated insofar as their financial knowledge levels (Bongini et al., 2016; De Vincentiis et al., 2017) and much less so about their attitudinal profile (Bocchialini & Ronchini, 2019).

By examining how financial self-efficacy beliefs, feelings and opinions towards finance, here labeled as "attitude towards finance", influence the students' financial knowledge in Italy, the present chapter seeks to address the above-mentioned research gaps and aims to answer these research questions: 1) To what extent do students in our sample possess financial knowledge and hold healthy attitudes towards finance? 2) Does attitude towards finance have a significant effect on financial knowledge?

Thus, it originally contributes to the body of knowledge in two main ways: by exploring the nature and magnitude of the relationship between attitudes towards finance and financial knowledge; by clarifying the direction of causality between them.

This chapter offers several contributions: (1) it adopts a different understanding of the key construct "attitude" in comparison with previous literature, that's also because the financial attitude construct has been recently questioned (D'Alessio et al., 2020); (2) the relationship between attitude towards finance and financial knowledge is here clarified in terms of both correlation and causation; and (3) the chapter specifies which of the three dimensions of attitude towards finance—vision, emotion, and self-confidence—leads most effectively to being financially knowledgeable (or vice versa). To our knowledge, no prior study in Italy or abroad has examined these aspects so far.

The remainder of this chapter is organized as follows. Section 6.2 presents a review of all studies that are relevant to our research questions. Section 6.3 describes the data and presents the methodology for investigating the linkages between attitude and knowledge. Section 6.4 discusses the empirical results. Finally, Sect. 6.5 summarizes the key findings and provides some ideas for further research.

6.2 Literature Review

The extant research in the financial education field has widely acknowledged financial attitude as an important driver of financial outcomes: attitudes can be regarded as a crucial factor to be considered when attempting to understand variability in one's financial literacy levels (OECD, 2019; Yahaya et al., 2019). Early studies mainly focused on measuring financial competencies and researching their determinants. Financial literacy has been found to be affected by a variety of factors such as education, age, gender, income, employment status, nationality, and family background (Atkinson & Messy, 2012; Chen & Volpe, 1998; Lusardi et al., 2010; OECD, 2013). Attitude has been shown to be also particularly important (Talwar et al., 2021).

Many studies on financial literacy have targeted young people so far and, in particular, the more educated ones attending high school, college, and university (Beal & Delpachitra, 2003; Chen & Volpe, 1998; de Bassa Scheresberg et al., 2014; Jones, 2005; Wagland & Taylor, 2009). In this general framework, both in Italy and abroad economics/business students have been frequently targeted by scholars to date (Bocchialini et al., 2013; Bongini et al., 2015; Gok & Ozkale, 2019; Kuntze et al., 2019). After all, this specific target population has some interesting features: this subgroup of millennials is better suited to explore the gender gap issue (Bongini et al., 2016) because it is characterized by homogeneity (especially in terms of educational choices). Moreover, given their range of age, university students are going through the delicate transition phase from dependency to financial independence from their parents. Business students also have higher financial exposure in comparison with younger students. Nevertheless, there is yet mixed evidence on whether economics/business students are financially educated or not. For example, Pintye and Kiss (2016) have shown that the financial literacy of economics and business Hungarian students—except in the dimension of financial behavior—cannot be considered to be at a higher level than among "average" young people. On the other hand, based on a sample of Italian university students, De Vincentiis, Pia, and Zocchi provide evidence that graduate students in Economics and Finance had above average (levels of) financial literacy compared to students graduates in different fields of study. Equally heterogeneous results come from the existing studies specifically aimed at assessing (inter alia) financial attitudes among university students (Setiyani & Solichatun, 2019; Yogasnumurti et al., 2020), probably due to the large variety of the definition and measurement methods used, which certainly do not favor comparisons over time and across countries.

The second strand of studies subsequently focused on measuring and researching mutual relationships between financial literacy, attitude (Ameliawati & Setiyani, 2018; Haque & Zulfiqar, 2015), and other significant outcomes, including, for example, economic empowerment and financial well-being (Consumer Financial Protection Bureau, 2015; Haque & Zulfiqar, 2016), financial satisfaction (Arifin, 2018), financial inclusion, and resilience (Dwivedi et al., 2015; Lyons et al., 2020). For example, (Skagerlund et al., 2018a, 2018b) have recently documented that some cognitive and emotional attitude towards numbers) are a driving force behind becoming financially literate.

Another strand of research in personal finances has explored possible inter-linkages among financial attitudes, financial knowledge, and financial behavior (Fessler et al., 2020; Rai et al., 2019; Yong et al., 2018). The general findings have shown that these variables are significantly and reciprocally interrelated (Kadoya & Khan, 2020; Shim et al., 2009). Gender differences in financial knowledge, attitude, and behavior have also been widely documented in various studies worldwide (Bucher-Koenen et al., 2017; Robson & Peetz, 2020).The specific relationship between financial knowledge financial behavior is another topic which has been extensively explored so far (Almenberg & Säve-Söderbergh, 2011; Van Rooij et al., 2012), whereas a smaller body of work has focused on both the attitudes-knowledge dyad (Borden et al., 2008; Jorgensen & Savla, 2010; Shim et al., 2010) and the attitude-behavior pair (Talwar et al., 2021). This is probably due to the more evanescent nature of the attitudes construct. In fact, it has been noted that it is far more difficult to determine what "sound attitudes" are, than to establish whether a behavior is healthy or an answer to a financial knowledge is correct (Johan et al., 2021; Riitsalu et al., 2019). Anyway, when it comes to the relationship between financial knowledge and financial attitude, the evidence is mixed: a positive relation has been documented in some studies (Hayhoe et al., 2005; Riitsalu et al., 2019) in the face of a null or a weak association found in others (Agarwalla et al., 2013; Riitsalu et al., 2019). Furthermore, to date, very few studies have focused on causation rather than correlation. Accordingly, further investigations are required in order to clarify the direction of causality between knowledge and attitude.

Yet, to date, some research has finally also investigated the actual impacts on financial knowledge exerted by each one of the three dimensions of the so-called attitude towards finance construct (Dobni & Racine, 2016; Driva et al., 2016). For example, in their recent study Van Der Cruijsen et al. (2021) found that a positive vision towards finance (approximated by the belief that the financial sector, its institutions, and regulators are trustworthy) is associated with a higher level of financial knowledge. Likewise, according to other prior studies, the feelings related to finance (for example, personal interest in financial matters rather than financial anxiety) as well as financial self-efficacy interferes with individuals' ability to attain financial knowledge (Arellano et al., 2014; Bongini et al., 2016; Farrell et al., 2016; Palameta et al., 2016; Skagerlund et al., 2018a, 2018b). Basically, despite the advantage of determining associations, these studies have investigated in isolation one single specific aspect of the multidimensional construct "attitude towards finance". To date, these factors have neither been explicitly seen as a dimension of the attitude towards finance nor related to the other components of the above construct within a single complete framework.

6.3 Data and Method

6.3.1 Sample

About 500 business students enrolled at the Department of Economics and Management of Parma University, in Northern Italy, voluntarily participated in this study. The survey has been administered during the first semester of the academic year 2019–2020. A total of about 600 questionnaires were distributed. After filtering, the final sample consisted of 466 respondents. Female respondents were slightly lower than male respondents (48.71% vs 51.29%). Participants were mostly of Italian nationality (89.70%), belonged to different academic years and different curricula. More than half of the sample were master's degree students (55.79%), whereas 44.21% were first-level degree students. When it comes to the type of degree course, about 36% of participants pursued the finance stream, whereas 64% pursued a degree in non-finance-related fields. Basically, the age of the respondents ranged from 19 and 29; the largest age group was between 23–25 years old (49.79%), followed by the age group 19–22 (30.04%). There were quite great differences in the area of origin from which participants came (49.36% were from the North of Italy, 36.27% from the South, 7.30% from the Center), as well as in their educational backgrounds and attainments at the school-leaving diploma. Just over half of the sample still lived with their families of origin (52.36%). An overview of the socio-demographic characteristics of the sample can be seen in Table 6.1.

6.3.2 *Methodology*

In order to assess both the level of the financial literacy of business students in the sample and their attitude profile, this research adopted a questionnaire method. The questionnaire consisted of three main sections covering: (1) socio-demographic information; (2) financial knowledge; and (3) attitude towards finance. The first part provided identification of respondents including their gender, age, nationality, area of origin, and level of study.

The second part of the questionnaire, aimed at measuring respondents' financial knowledge level, was based on six questions, based on prior literature (Lusardi & Mitchell, 2011; Van Rooij et al., 2012). They covered several topics such as compound interest, inflation, stock risk, investment risk assessment, the relationship between interest rate and price of bonds, and portfolio diversification. For each question, 1 point was awarded for the correct answer and 0 points for the wrong or missing answer. A dummy variable was created indicating if the question was answered correctly. Thus, participants' financial knowledge total score (FK_tot) was equal to the sum of their correct answers and ranged from 0 to 6.

Finally, the last part of the questionnaire gauged respondents' attitude profiles. The test consisted of 51 statements adapted from the study

Table 6.1 Study participant characteristics

		Percent (%)
Gender	Male	51.29
	Female	48.71
Age	19–22	30.04
	23–25	49.79
	26–29	18.45
	From 30 upwards	1.72
Nationality	Italian	89.70
	Foreign	10.20
Area/Region of origin	North	49.36
	Center	7.30
	South	36.27
	Foreign	7.08
Level of University study	First Cycle	44.21
	Second Cycle	55.79
Type of Degree course	Finance	35.84
	No Finance	64.16
Type of High school diploma	Scientific High School or Similar	39.70
	Classical Or Linguistic or Humanistic/Social High School	16.95
	Technical Commercial Institute	36.05
	Technical-Industrial or Tourist or Hotel-Management Institute	3.00
	Institute For Surveyors	1.50
	Professional Institute	2.79
High school mark	60–65	7.94
	66–70	12.66
	71–75	16.74
	76–80	16.52
	81–85	12.02
	86–90	13.52
	91–95	6.22
	96–100	14.38
Cohabitation	Off-site students (student living alone/away from family)	47.21
	Students living with family	52.36

of Bocchialini and Ronchini (2019), taking also into account the "Attitude towards Economics" questionnaire (Walstad & Soper, 1983). The statement focused on the three different facets of the attitude towards finance construct, namely: thoughts and beliefs related to "finance", emotional disposition towards finance, and financial self-efficacy beliefs. Participants were asked to indicate the degree to which they agreed about the following items: (1) finance is a difficult and math-heavy subject; (2) financial education is useful in their daily and professional life; (3) finance is a male domain; and (4) financial skills are fixed (namely not malleable). Participants were also asked (5) how they emotionally feel when dealing with finance-related issues and (7) to rate their level of self-confidence in financial matters. Like in Pisa 2015, all questionnaire statements utilized a four-point Likert scale, ranging from "strongly disagree", "somewhat disagree", and "somewhat agree" to "strongly agree". In more detail, vision towards finance was measured using 30-items, emotional disposition towards finance was measured with 9-items, and self-efficacy beliefs were measured using 12-items. Accordingly, three different indicators, one for each dimension of attitude, were created (the "VIEW" indicator; the "EMOTION" and the "SELF-CONF" indicator) and next converted into a standardized score on a basis of 1 for comparison. Basically, the overall "attitude towards finance (Potrich et al., 2016).

In this chapter, the structural equation modeling (SEM) approach was employed to test our hypotheses about causal relationships between attitude towards finance and financial knowledge, based on the theoretical background and the research questions discussed above. In order to investigate whether a good/bad attitude to finance causes high/poor financial knowledge, or whether the relationship is the other way around, we applied SEM to the collected data. The causal relationship between the attitude to finance and actual financial knowledge was estimated. We also studied which one of the different attitude facets was most closely related to students' financial knowledge. Finally, we identified the most significant exogenous variables provided by the questionnaire.

Hence, the research variables in the study were as follows: the attitude towards finance variable was proxied from Bocchialini and Ronchini (2019) with the view of finance, emotional disposition towards finance, and self-efficacy beliefs (or perceived competence in finance) sub-indicators. The financial knowledge variable was divided into the basic financial knowledge indicator and the advanced financial knowledge indicator, respectively, proxied from Lusardi and Mitchell (2011) and Van Rooij et al. (2012). Their sum gave the overall indicator of financial knowledge.

Widely used in most behavioral, educational, medical, and social studies, SEM is a covariance-based statistical methodology able to capture causality relations between variables which can be either measurable (manifest variables) or not measurable (latent variables or factors) (Bollen, 1989; Kaplan, 2009). A variable is not measurable when the values assigned to it are uncertain because, for example, of errors generated by the measurement method. SEM can be divided into two sub-models: (1) structural or internal models which capture relationships between latent variables; (2) measure or external models which capture relationships between manifest and latent variables.

The relations between variables can be estimated using either covariance-based methods or component-based methods. Covariance-based methods work mainly on manifest variables, and component-based methods on latent variables, through multivariate linear techniques, in particular path analysis, which was introduced by Wright, in 1921 to genetics research and originally applied by Joreskog (1973) in SEM. The underlying mathematical tool is the decomposition of the total correlation or covariance between two variables among all paths which connect them. Decomposition of total correlation produces the path coefficients, which express the strength of the causality relation. The path diagram is the graphical representation of a system of the simultaneous equations where latent and manifest variables are represented by circles and squares respectively.

SEM has been used to study attitude to different subjects: in particular, in mathematics (Papanastasiou, 2000; Yurt, 2014), in statistics (Escalera-Chávez et al., 2014), and in finance (Nadeem et al., 2020; Potrich et al., 2016; Talwar et al., 2021).

Based on these studies, the next section presents the relationship between attitude towards finance and financial knowledge. It should be noted that for all the SEM analyses described, we monitored the most relevant indicators for the goodness of the model: the Goodness of Fit Index, the Standardized Root Mean Square Residual, the baseline model Chi-Square, and the Satorra-Bentler-Scaled Base Model Chi-Square. We also verified whether they were above or below the suggested thresholds (e.g., the Goodness of Fit Index above 0.9, the Standardized Root Mean Square Residual below 0.08). All the indicators of model fit were found within the permitted levels and were consistent with past studies (Rai et al., 2019).

Table 6.2 shows and describes the dependent and independent variables used in the analysis, which was conducted using SAS (https://www.sas.com/). For other possible statistical packages, see Narayanan (2012).

Table 6.2 Description of variables used in this study

Variables	*Abbreviation*	*Description*
Socio-demographic variables		
Gender	GEN	Student gender. Dummy variable (1 = Male, 0 = Female)
Age	AGE	Student age
Nationality	NAT	Student country of origin. Dummy variable (1 = Italian, 0 = Other)
Area of origin	AREA	Student region of origin. (NORTH, CENTRAL, SOUTH, and FOREIGN, according to the classification of regions by ISTAT). The appropriate dummy variables were constructed
High school diploma	HSD	Student diploma type (scientific high school or similar, classical or linguistic or humanistic/social high school, technical-commercial institute, technical-industrial or tourist or hotel-management institute, institute for surveyors, professional institute). The appropriate dummy variables were constructed
Mark of high school diploma	HSD_M	Students' final diploma result (60–65; 66–70; 71–75; 76–80; 81–85; 86–90; 91–95; 96–100)
Mathematics	MATH	Self-assessment of liking mathematics (Likert-type scale 1–4—None, very little, some, lot)

(continued)

Table 6.2 (continued)

Variables	*Abbreviation*	*Description*
Study levels	LEVEL	Student level of university education. Dummy variable (1 = First level/bachelor's degree; 0 = Master's degree/second level)
Student Year	YEARS	Year of the student (First level: first year, second year, third year, outside prescribed time for the bachelor's degree; Second level: first year, second year, outside prescribed time for the master's degree)
Father's educational attainment	FATHER_EDU	Father's Education (No schooling, completed primary school, completed junior high school a, completed vocational school, Completed high school, University graduate)
Mother's educational attainment	MOTHER_EDU	Mother's Education (No schooling, completed primary school, completed junior high school, completed vocational school, Completed high school, University graduate)
Household income	INCOME	Income level of the household
Ownership of material goods	Mobile TV Pc_tablet Car Bathroom Boat Moto Hsea Hmountain	Self-assessment of household possess: 0, 1, 2, 3, or more (the material goods were mobile phone, television, personal computer-tablet, car, bathroom, private boat, motorbike, beach house, mountain house)
Financial Knowledge *indicator–discrete variable*		
Indicator of total financial knowledge	FK_tot	Indicator of the student's overall financial knowledge (sum of the correct answer to the six questions in the category: 1 = Correct, 0 = Incorrect/Missing)

(continued)

Table 6.2 (continued)

Variables	*Abbreviation*	*Description*
Attitude towards finance indicator–discrete variable		
"View of finance" component	VIEW	Indicator of ATF, with specific regard to the component "view of finance". Self-assessment of vision towards finance (sum of the scores on 30 statements—Likert-type scale 1–4). The index is normalized and returned to a unit basis
"Emotional disposition towards finance" component	EMOTION	Indicator of ATF, with specific regard to the component "emotional disposition towards finance". Self-assessment of feelings towards finance (sum of the scores on 9 statements—Likert-type scale 1–4). The score is normalized
"Self-confidence" component	SELF-CONF	Indicator of ATF, with specific regard to the component "self-confidence towards finance". Self-assessment of financial self-efficacy beliefs (sum of the scores on 12 statements—Likert-type scale 1–4). The score is normalized
Overall Indicator of Attitude Towards Finance	ATF	Indicator of the student's overall attitude towards finance, derived from the sum of the previous three indicators ("VIEW" indicator + "EMOTION" indicator + "SELF-CONF" indicator)

6.4 Results

6.4.1 *Students' Levels of Financial Knowledge and Attitude Towards Finance*

The first objective of this chapter was to assess the levels of financial knowledge and attitude towards finance of the students in the sample. Other than the overall measure of financial knowledge, we also provided the measure for two subscales—basic financial knowledge and sophisticated financial knowledge. Tables 6.3 and 6.4, respectively, show the main descriptive statistics for each of the financial knowledge indexes described above and their frequency distributions.

Based on the results exhibited in Tables 6.3 and 6.4, it is known that, on average, the general performance in financial knowledge is relatively good among university students in the sample. Only about 24% of respondents get all six correct and approximately 7% of students fall into the cluster "zero correct answers-to-1 correct answers"; the overall measure of financial knowledge of the sample can be considered medium–high.

When it comes to basic financial knowledge, more than one in two students in the sample were top performers (51.50%). They obtained the maximum score of 3 and thus demonstrated to fully possess the set of knowledge that underpins day-to-day financial decision-making, given that they correctly responded to questions about interest rates, inflation, and risk diversification. For example, 82.00% of the respondents know how compound interest is calculated, about 70% of them understand the concepts of inflation as well as the risk diversification principle. For the advanced knowledge indicator, 33.38% of the sample obtained a maximum score of 3. These students demonstrated to possess some complex financial concepts and showed a deep understanding of the financial landscape. Interestingly, students scored the highest on question Q5, which evaluated the knowledge of the differences between

Table 6.3 Descriptive statistics of FK indicators

Variable	*Obs*	*Mean*	*Mode*	*Median*	*Std. Dev*	*Min*	*Max*
FK_tot	466	4.217	5	5	1.560	0	6

Table 6.4 Frequency distribution of FK_tot indicator

FK_tot	Score	Frequency (%)
	0	13 (2.79)
	1	20 (4.29)
	2	36 (7.73)
	3	59 (12.66)
	4	103 (22.10)
	5	126 (27.04)
	6	109 (23.39)

stocks and bonds, followed by Q. 1 (interest rates) and Q. 3 (risk diversification). Conversely, the lowest scores were obtained on question Q4, which deals with the relationship between bond prices and interest rates, followed by Q. 2 on the understanding of inflation.

The attitudinal profiles towards finance were also assessed to consider some personality factors, such as cognitive and emotional status towards finance, which could play an important role in affecting the financial knowledge of students in the sample. Table 6.5 reports the descriptive statistics of the attitude towards finance (ATF) of the sample. The frequency distribution for the "attitude towards finance" measures are presented in Tables 6.6a and 6.6b.

Looking at them, it can be noted that for none of the indicators the descriptive statistics of mean/median/mode fall below the value of 0.7, showing a highly positive average level of attitude towards finance in the sample. The overall attitude towards finance index, as well as its subcomponents, were quite high: for each attitude subcomponent, about 70% of the sample were in the high and very high levels. Only a minority of students (less than 5% of the sample) had a strong disengagement with finance and was categorized as very low or low profile. Moreover, since the proportion of students with a positive attitude towards

Table 6.5 Descriptive statistics of "attitude towards finance" indicators

ATF Indicators	*Obs*	*Mean*	*Median*	*Mode*	*Min*	*Max*	*Range*
View	466	0.737	0.767	0.833	0	1	0–1
Emotion	466	0.764	0.778	0.833	0	1	0–1
Self-Confidence	466	0.715	0.729	0.854	0	1	0–1
Overall ATF	466	0.736	0.752	0.819	0	1	0–1

Table 6.6a Frequency distribution of "attitude towards finance" overall indicator

Level		*Overall ATF Frequency (%)*
Very Low	0–0.20	6 (1.29)
Low	0.21–0.40	4 (0.86)
Medium	0.41–0.60	45 (9.66)
High	0.61–0.80	260 (55.79)
Very High	0.81–1	151 (32.40)

Table 6.6b Frequency distribution of "attitude towards finance" indicators (View–Emotion–Self-Confidence)

Level		*View* *Frequency (%)*	*Emotion* *Frequency (%)*	*Self-Confidence* *Frequency (%)*
Very Low	0–0.20	8 (1.72)	1 (0.21)	8 (1.72)
Low	0.21–0.40	3 (0.64)	15 (3.22)	10 (2.15)
Medium	0.41–0.60	31 (6.87)	58 (12.45)	98 (21.03)
High	0.6–0.80	293 (62.87)	186 (39.91)	189 (40.56)
Very High	0.81–1	130 (27.90)	206 (44.21)	161 (34.55)

finance was large within the sample, some heterogeneity anyway existed related to the single sub-components of the construct (vision, emotion and self-confidence). In fact, there existed different "profiles of attitude" within the sample, depending on the mix of a personal vision of finance, the beliefs about the self, and the emotional reaction to financial issues held by each student. In particular, the differences mainly concerned the emotional scores and the "beliefs about the self" component levels.

In order to better understand the attitude profiles of the participants we can then apply this rule: we consider (1) the individual attitude profile positive if all three of its constituent components were positive, while (2) we judge the attitude profile negative if only one of its constituent dimensions was rated as negative. Following this rule, we find that 84.76% of the sample has a positive attitude towards finance, while 15.24% has a negative attitude (because at least one of the components was negative, i.e., below a score of 0.5).

6.5 Relationship Between Attitude Towards Finance and Financial Knowledge

As noted above, prior investigations explored the link between financial attitude and financial knowledge, but the evidence was mixed (Agarwalla et al., 2013; Hayhoe et al., 2005; Riitsalu et al., 2019). Moreover, none of those studies focused on the construct of attitude towards finance here investigated and the causality question rested to be unraveled as well.

To identify which way the causation runs between attitude and knowledge, we used two sets of SEM. The first takes the attitude towards finance as the independent variable which causes knowledge (ATF ->

FK), whereas the second takes the knowledge as the independent variable which affect the attitude towards finance (as FK -> ATF). The two sets of models are run on the questionnaire data and the *p*-values of the model parameters were compared: the model with the lowest p-values can be considered the best.

Table 6.7 summarizes the *p*-values of the main parameters of a model ATF -> FK_tot and FK_tot -> ATF, where FK_tot is the financial knowledge indicator. Comparing the pairs of *p*-values present in each row, it was clear that the parameters of the financial attitude in the model ATF -> FK_tot were generally smaller than the parameters of the attitude towards finance in a model FK_tot -> ATF. Namely, this was the case for both View and Emotion while for Self Confidence the two *p*-values were rather close. Thus, the first model was preferable.

Based on Table 6.7, it was found that the levels of all *p*-values were quite low; this suggested that a direct and significant link existed between attitude and knowledge (the smaller the *p*-values in a study, the more the null hypothesis is improbable, and the alternative hypothesis is probable). Moreover, the direction of the causation was also suggested: attitude towards finance affected financial knowledge rather than the opposite; the more favorable a person's profile of ATF, the higher the level of Financial Knowledge.

After having established that the strongest causality relation was from attitude to knowledge, we focused on SEM where attitude towards finance was the dependent/response variable and financial knowledge was the independent/exogenous variable. Now we aimed at investigating

Table 6.7 *p*-values obtained in a SEM model where it is assumed that (1) each "ATF" component (first column) determines the financial knowledge "FK" (column ATF -> FK) and (2) the financial knowledge "FK" affect each "ATF" components (column FK -> ATF)

ATF dimensions	*Overall financial knowledge*	
	ATF -> FK_tot	*FK_tot -> ATF*
View	2.89E-7	7.25E-6
Self Confidence	2.68E-12	2.28E-12
Emotion	5.62E-12	2.79–11

which dimension of attitude towards finance most influenced the knowledge of finance and thus played the most significant role in becoming financially knowledgeable.

The three components of attitude—view of finance, perceived financial competence, and emotional disposition—were the result of aggregation of the scores obtained in many statements (respectively, 30, 12, and 9 for each component). As all statements didn't have the same relevance on basic and advanced financial knowledge indicators, we therefore build 30 + 12 + 9 = 51 SEM models to study the significance of the parameters of each relation and re-compute the three attitude components using only the statements for which the relation with the two FK indicators was both significant at 0.01 level. Of the 30 statements related to the view of finance, only 8 were significant and were retained; 7 (out of 12) were retained concerning the perceived financial competence; and 7 (out of 9) questions relating to emotional disposition.

The results of the SEM obtained using the re-computed attitude components are in Table 6.8. Looking at the extent to which each dimension of attitude was associated with financial knowledge in the sample, the results revealed that all three components of attitude had a positive impact on the overall indicator of knowledge, with an emotional disposition towards financial matters exerting the strongest influence, followed by financial self-confidence. Instead, we found that the "view towards finance" component of the overall attitude towards finance index exerted the lowest effect on the general index of financial knowledge.

Next, we tested the extent to which the three components of attitude were interlinked. Table 6.9 reported the results: the three re-computed dimensions of attitude were positively associated with each other. Financial self-confidence and emotional disposition towards finance were the most strongly associated, with a Pearson coefficient larger than 0.8 congruent with a previous study (Lind et al., 2020). Thus, our result

Table 6.8 *p*-values obtained in SEM where scores for "view", "self-confidence", and "emotion" are re-computed using only significant statements

	ATF -> FK_tot
View	2.32eE-8
Self Confidence	6.18E-10
Emotion	7.29E-12

Table 6.9 Statistical associations between the three dimensions of attitude: Pearson Coefficients

	View	*Self confidence*	*Emotion*
View	1	0.67450	0.61641
Self Confidence		1	0.82424
Emotion			1

suggests the complementary nature and the close interplay between emotional disposition towards finance and students' self-efficacy beliefs in finance. Accordingly, a negative emotional disposition towards finance (for example in the shape of financial anxiety) can be the reason for a lack of success/ understanding, because it is well known that anxious people have access to fewer cognitive resources. At the same time, a low perceived competence related to finance tend to elicit a negative emotional disposition towards finance-related situations. Conversely, if finance generates positive emotions in an individual, he or she will feel comfortable and benign in presiding over finance and vice versa.

6.6 Discussion and Conclusion

Globally, financial knowledge is now recognized as a key basic skill and a "buffer against adversities". It serves not only to navigate in modern society in normal times but also for promoting financial resilience by lowering vulnerabilities in difficult times and in complex environmental scenarios, such as the current one (Lyons et al., 2020). Many adversities and threats—whether environmental, social, or economic—particularly hang over young adults and the younger. In Italy, where both the NEET and the financial illiteracy rates are historically high (see also Chapter 6.1), young adults have been severely impacted also by the latest economic crisis making harder their transition to work. In addition, at present, they are also experiencing a significant economic impact on the current coronavirus pandemic.

In this analysis, we addressed the "young adults' issue" by exploring the mechanisms through which the attitudinal traits influence their achievement in the financial knowledge test. In other words, we wanted to know (1) whether the attitude towards finance (namely, their beliefs to finance, feelings, financial self-confidence) matters when it comes to financial knowledge and (2) how the attitude towards finance-financial knowledge circle unfolds. In doing so, we intended to measure attitude towards finance in a sample of Italian university students and to assess their financial knowledge, using the definitions outlined in the introduction. We also explored the relationship between attitude towards finance-financial knowledge in terms of both correlation and causation. Prior works in the financial field aimed at deepening these themes have yielded quite inconclusive results and, in any case, they have focused on constructs other than the one investigated here, constructs that have also been criticized recently (D'Alessio et al., 2020).

We found that the university students surveyed showed both a relatively high level of financial knowledge and a quite positive attitude towards finance; only a small minority of them had a low profile in the area of attitude. Our findings are consistent with other studies of financial knowledge of college student populations (Anderson et al., 2018; Kubicková et al., 2019). Moreover, our perspective substantiates and also widens previous findings on the effects on financial knowledge exerted by certain variables, such as the interest in financial matters or financial self-confidence (Arellano et al., 2018; Bongini et al., 2016; Bucher-Koenen et al., 2017; Grohmann, 2016). For the first time, the present study has included these elements in the realm of attitude, rather than considering them as separate domains.

It was found that a direct and significant correlation exists between attitude towards finance and financial knowledge and that the former does affect the latter. It follows that a positive attitude towards finance can be a good start in becoming a financially knowledgeable person. This result has important implications for policymakers: the attitude towards finance should be targeted to enhance financial knowledge among the young. In order to optimize learners' financial outcomes, financial literacy programs should develop specific criteria for diagnosing and eventually modifying attitudes. In fact, learners who enter a financial literacy program with positive or even neutral attitudes towards finance are more likely to be open and willing to learn about finance; at the same time, favorable attitudes to finance should represent a final goal of a training program. Consequently, in agreement with Zan and Di Martino (2007), we believe that it is important to carefully monitor the learners' attitude profile, especially in the case of "negative" attitudes, because "the diagnosis of a 'negative' attitude becomes a starting point to design an intervention aimed at modifying the component(s) identified as 'negative'". It follows that possible remedial actions aimed at pushing a "positive attitude" will require a completely different approach depending on whether the negative attitude to restore refers only to the emotional component or it refers to a particular pattern of (distorted and wrong) beliefs and emotions.

This study has also determined which of the three components of the attitude towards finance construct played the main role in becoming financially knowledgeable. It was found that the emotional disposition towards financial matters was the strongest predictor of financial knowledge, followed by financial self-confidence and next by the view of finance. In fact, the most financially knowledgeable students in our sample were more likely to be personally interested in financial learning (or other finance-related experiences) and to believe that they could easily succeed at learning finance. Our results were in line with prior studies, which found that feelings relate to finance and financial self-efficacy impacts an individuals' ability to gain financial knowledge (Arellano et al., 2014; Farrell et al., 2016; Skagerlund et al., 2018a, 2018b). Additionally, we have also pointed out how the way emotion relates to finance is strictly associated and interacts with the financial self-efficacy beliefs. Therefore, discouraging negative feelings towards finance should have interesting repercussions on low self-efficacy beliefs too and vice versa; these actions can support the learners' financial literacy process.

This research has certain limitations. The first and most important is that we collected data through a self-reported questionnaire at one point in time, in a single campus and geography. The small size of the non-random sample doesn't allow any generalization to a larger population beyond the scope of the present study. Thus, further research needs to be done to replicate our findings targeting a different subgroup of the population in Italy or in other countries. Anyway, more studies in the finance framework need to focus on the attitude towards finance-financial knowledge relations to validate our findings on a larger scale; the causal link between them could also be explored by using an experimental design. Future research should even assess the influence of some socio-demographic factors (age, gender, math aptitude, parents' attitude towards finance…) here neglected to affect and determine the level of students' financial knowledge and attitude. Future investigations are also required to explore whether the so-called attitude towards finance construct may also affect financial behavior and well-being, which are the true goals of financial literacy initiatives. Finally, exploring how to develop the right attitude towards finance through formal financial education programs is another interesting avenue for further study. Worldwide there is still much to be done to integrate the "attitude diagnosis" into effective financial literacy curricula.

References

Agarwalla, S. K., Barua, S. K., Jacob, J., & Varma, J. R. (2013). Financial literacy among working young in urban India. *Indian Institute of Management Ahmedabad, 10*, 2. https://doi.org/10.2139/ssrn.2338723

Aina, C., Brunetti, I., Mussida, C., & Scicchitano, S. (2021). *Even more discouraged? The NEET generation at the age of Covid19*. ISSN 27848701. https://doi.org/10.53223/InappWP_2021-68

Almenberg, J., & Widmark, O. (2011a). *Numeracy and financial literacy among Swedish adults* (Report). Swedish Financial Supervisory Authority, Finansinspektionen.

Almenberg, J., & Widmark, O. (2011b). Numeracy, financial literacy and participation in asset markets. SSRN. https://doi.org/10.2139/ssrn.1756674

Ameliawati, M., & Setiyani, R. (2018). The influence of financial attitude, financial socialization, and financial experience to financial management behavior with financial literacy as the mediation variable. *ICE-BEES 2018 International Conference on Economics, Business and Economic Education.*

Anderson, Drew M., Conzelmann, J., & Lacy, T. A. (2018). *The state of financial knowledge in college: New evidence from a national survey*. Santa Monica, CA: RAND Corporation. https://www.rand.org/pubs/working_papers/WR1256.html

Arellano, C., Cámara, N., & Tuesta, D. (2018). Explaining the gender gap in financial literacy: The role of non-cognitive skills. *Economic Notes: Review of Banking, Finance and Monetary Economics, 47*(2–3), 495–518.

Arellano, A., Cámara, N., & Tuesta, D. (2014). *The effect of self-confidence on financial literacy* (BBVA Research Working Paper, No. 14/28). Fundación BBVA, October, Madrid.

Arifin, A. Z. (2018). Influence factors toward financial satisfaction with financial behavior as intervening variable on Jakarta area workforce. *European Research Studies Journal, 21*(1), 90–103.

Atkinson, A., & Messy, F. (2012). *Measuring financial literacy results of the oecd/international network on financial education (INFE)*. Pilot Study.

Beal, D. J., & Delpachitra, S. B. (2003). Financial literacy among Australian university students. *Journal of Applied Economics and Policy, 22*(1), 65–78.

Bocchialini, E., Milioli, M. A., Poletti, L., & Ronchini, B. (2013). Lo stato dell'arte dell'educazione finanziaria tra gli studenti universitari in Italia. *BANCARIA, 12*, 30–48.

Bocchialini, E., & Ronchini, B. (2019). A pilot study assessing attitudes toward finance among Italian business students. *International Journal of Business and Management, 14*(10). https://doi.org/10.5539/ijbm.v14n10p44

Bollen, K. A. (1989). *Structural equations with latent variables*. John Wiley & Sons.

Bongini, P., Trivellato, P., & Zenga, M. (2016). Financial literacy and undergraduates: Application of latent regression rasch model. *Journal of Financial Management, Markets and Institutions*, 1, 23–42.

Bongini, P., Colombo, L., & Iwanicz Drozdowska, M. (2015). Financial literacy: Where do we stand? *Journal of Financial Management, Markets and Institutions, 1*. https://doi.org/10.12831/80527

Borden, L. M., Lee, S.-A., Serido, J., & Collins, D. (2008). Changing college students' financial knowledge, attitudes, and behavior through seminar participation. *Journal of Family and Economic Issues, 29*, 23–40.

Bucher-Koenen, T., Lusardi, A., Alessie, R., & van Rooij M. (2017). How financially literate are women? An overview and new insights. *Journal of Consumer Affairs, 51*(2), 255–283. https://doi.org/10.1111/joca.12121

Chen, H., & Volpe, R. P. (1998). An analysis of personal financial literacy among college students. *Financial Services Review, 7*, 107–128.

Bureau, C. F. P. (2015). *Financial well-being: The goal of financial education*. Consumer Financial Protection Bureau.

D'Alessio G., De Bonis, R., Neri, A., & Rampazzi, C. (2020). L'alfabetizzazione finanziaria degli italiani: i risultati dell'indagine della Banca d'Italia del 2020, Questioni di Economia e Finanza (Occasional Papers) n. 588. https://doi.org/10.2139/ssrn.3826435

Danes, S., & Haberman, H. (2007). Teen financial knowledge, self-efficacy, and behavior: A gender view. *Journal of Financial Counselling and Planning, 18*(2), 48–60.

De Bassa Scheresberg, C., Lusardi A., & Yakoboski, P. J. (2014). *College-educated Millennials: An Overview of their personal finances*. TIAA-CREF Institute.

De Vincentiis, P., Pia, P., & Zocchi, P. (2017). Conoscenze finanziarie e overconfidence dei giovani laureate. *Bancaria*, (2).

Di Martino, P., & Zan, R. (2011). Attitude towards mathematics: A bridge between beliefs and emotions. *ZDM: The International Journal On Mathematics Education, 43*(4), 471–482. https://doi.org/10.1007/s11858-011-0309-6

Dobni, D. M., & Racine, M. D. (2016). Investors' image of the stock market: Antecedents and consequences. *Financial Services Review, 25*, 1–28.

Driva, A., Lührmann, M., & Winter, J. (2016). Gender differences and stereotypes in financial literacy: Off to an early start. *Economics Letters, 146*, 143–146. https://doi.org/10.1016/j.econlet.2016.07.029

Dwivedi, M., Purohit, H., & Mehta, D. (2015). Improving financial literacy among women: The role of universities. *Economic Challenger*, (October–December). https://ssrn.com/abstract=2668034

Escalera-Chávez, M. E., García-Santillán, A., & Venegas-Martínez, F. (2014). Modeling attitude toward statistics by a structural equation. *Eurasia Journal of Mathematics, Science and Technology Education, 10*(1), 23–31.

Farrell, L., Fry, T., & Risse. L. (2016). The significance of financial self-efficacy in explaining women's personal finance behaviour. *Journal of Economic Psychology, 54*(C), 85–99.

Fessler P., Silgoner M., & Weber R. (2020). Financial knowledge, attitude and behavior: Evidence from the Austrian Survey of Financial Literacy. *Empirica, Springer; Austrian Institute for Economic Research; Austrian Economic Association, 47*(4), 929–947.

Garber, G., & Koyama, S. M. (2016). *Policy-effective financial knowledge and attitude factors* (Working Papers Series, 430). Central Bank of Brazil, Research Department.

Gok, I. Y., & Ozkale, A. (2019). Testing the influence of college education on the financial literacy level of university students in Turkey. *e-Journal of Business Education & Scholarship of Teaching, 13*(1), 46–58.

Grohmann, A. (2016). The gender gap in financial literacy: Income, education, and experience offer only partial explanations. *DIW Economic Bulletin, 6*(46/47), 531–537.

Haque, A., & Zulfiqar M. (2016). Women's economic empowerment through financial literacy, financial attitude and financial wellbeing. *International Journal of Business and Social Science, 7*(3).

Hayhoe, C., L., Leach, M. A., & Edwards, R. (2005). Credit cards held by college students. *Financial Counseling and Planning, 16*, 1–10. https://doi.org/10.1002/9781118619179

Huston, S. (2010). Measuring financial literacy. *Journal of Consumer Affairs, 44*(2), 296–316.

Johan, I., Rowlingson, K., & Appleyard, L. (2021). The effect of personal finance education on the financial knowledge, attitudes and behaviour of university students in Indonesia. *Journal of Family and Economic Issues 42*, 351–367. https://doi.org/10.1007/s10834-020-09721-9

Jones, J. E. (2005). College student knowledge and use of credit. *Journal of Financial Counseling and Planning, 16*(2), 9–16.

Jorgensen B. L., & Savla, J. (2010). "Financial literacy of young adults: The importance of parental socialization." *Family Relations, 59*(4), 465–478. https://doi.org/10.1111/j.1741-3729.2010.00616.x

Kadoya, Y., & Khan, M. S. R. (2020). Financial literacy in Japan: New evidence using financial knowledge, behavior, and attitude. *Sustainability, 12*(9), 3683. https://doi.org/10.3390/su12093683

Kaplan, D. (2009). *Structural equation modeling: Foundations and extensions* (2nd ed.). SAGE Publications. https://doi.org/10.4135/9781452226576

Kubicková, D., Nulicek, V., & Jindrichovska, I. (2019). "Is the financial literacy affected by the field of study? (A comparison of specialised secondary schools)." *International Journal of Management Science and Business Administration*, *5*(6), 15–20. Inovatus Services Ltd.

Kuntze M., Wu, C., Wooldridge, B. R., & Whang, Y. (2019). Improving financial literacy in college of business students: Modernizing delivery tools. *International Journal of Bank Marketing*, *37*(4), 976–999.

Lusardi A., & Mitchell O. (2011), Financial literacy around the world: An ovierview (NBER Working Paper 17107O).

Lusardi, A., Mitchell, O. S., & Curto, V. (2010). Financial literacy among the young. *Journal of Consumer Affairs*, *44*(2), 358–380.

Lusardi, A. (2015). Financial literacy skills for the 21st century: Evidence from PISA. *The Journal of Consumer Affairs*, *49*(3), 639–659. https://doi.org/10.1111/joca.12099

Lyons, A. C., Kass-Hanna, J., Liu, F., Greenlee, A. J., & Zeng, L. (2020). Building financial resilience through financial and digital literacy in South Asia and Sub-Saharan Africa (ADBI Working Paper 1098). Tokyo: Asian Development Bank Institute. https://www.adb.org/publications/building-financial-resilience-through-financial-digitalliteracy-south-asia-saharan-africa

Nadeem M. A., Qamar, M. A. J., Nazir, M. S., Ahmad, I., Timoshin, A., & Shehzad, K. (2020). How Investors attitudes shape stock market participation in the presence of financial self-efficacy. *Frontiers in Psychology*, *11*(22), 553351. https://doi.org/10.3389/fpsyg.2020.553351. PMID: 33192804; PMCID: PMC7642217.

Narayanan, A. (2012). A review of eight software packages for structural equation modelling. *The American Statistician*. https://doi.org/10.1080/00031305.2012.708641

Nicolini, G., (2019). Financial literacy in Europe. Assessment methodologies and evidence from european countries. Routledge. https://doi.org/10.4324/9780429431968

Obagbuwa, O., Akande, J., & Dev Tewari, D. (2021). Acquired financial knowledge and attitudes of South African university students and their financial decisions. *The Journal of Accounting and Management*, *1*(11), 96–110.

OECD. (2013). *OECD skills outlook 2013: First results from the survey of adult skills*. OECD Publishing.

OECD. (2015). *OECD/INFE toolkit for measuring financial literacy and financial inclusion*.

OECD. (2020). *OECD/INFE 2020 International survey of adult financial literacy* www.oecd.org/financial/education/launchoftheoecdinfeglobalfinancialliteracysurveyreport.htm

OECD INFE. (2011). Measuring financial literacy: Core questionnaire in measuring financial literacy: Questionnaire and guidance notes for conducting an internationally comparable survey of financial literacy. OECD.

OECD. (2019). *PISA 2018 assessment and analytical framework, PISA*. OECD Publishing.

Palameta, B., Nguyen, C., Hui, T. S., & Gyarmati, D. (2016). *The link between financial confidence and financial outcomes among working-aged Canadians*. The Social Research and Demonstration Corporation.

Papanastasiou, C. (2000). Effects of attitudes and beliefs on mathematics achievement. *Studies in Educational Evaluation, 26*, 27–42. https://doi.org/10.1016/S0191-491X(00)00004-3

Pintye, A., & Kiss, M. (2016). Financial literacy of students in business and economics higher education. *Annals of the University of Oradea, Economic Science Series, 25*(1), 781–789.

Potrich A. C. G., Vieira K. M., & Mendes-Da-Silva W. (2016). Development of a financial literacy model for university students. *Management Research Review, 39*(3), 356–376. https://doi.org/10.1108/MRR-06-2014-0143

Rai, K., Dua, S., & Yadav, M. (2019). Association of financial attitude, financial behaviour and Financial knowledge towards financial literacy: A structural equation modeling approach. *FIIB Business Review, 8*(1), 51–60. https://doi.org/10.1177/2319714519826651

Remund, D. L. (2010). Financial literacy explicated: The case for a clearer definition in an increasingly complex economy. *Journal of Consumer Affairs*. https://doi.org/10.1111/j.1745-6606.2010.01169.x

Riitsalu L., Murakas R., & Veeret D. (2019). Disentangling financial literacy: Three-dimensional approach to analysing management of personal finances in Estonia. *Journal of Social Research and Policy*, (April).

Robson, J., & Peetz, J. (2020). Gender differences in financial knowledge, attitudes, and behaviors: Accounting for socioeconomic disparities and psychological traits. *Journal of Consumer Affairs*, 54(3), 813–835. https://doi.org/10.1111/joca.12304

Setiyani R., & Solichatun, I. (2019), Financial well-being of college students: An empirical study on mediation effect of financial behavior. *KnE Social Sciences, 3*(11), 451–474. https://doi.org/10.18502/kss.v3i11.402

Shim, S., Barber, B. L., Card, N. A., Xiao, J. J., & Serido, J. (2010). Financial socialization of first- year college students: The role of parents, work, and education. *Journal of Youth and Adolescence, 39*(12), 1457–1470.

Shim, S., Xiao, J. J., Barber, B. L., & Lyons, A. C. (2009). Pathways to Life Success: A conceptual Model of Financial Well-being for Young Adults. *Journal of Applied Developmental Psychology, 30*(6), 708–723.

Skagerlund, K., Lind, T., Strömbäck, C., Tinghög, G., & Västfjäll, D. (2018a). Financial literacy and the role of numeracy–How individuals' attitude and affinity with numbers influence financial literacy. *Journal of Behavioral and Experimental Economics, 74*, 18–25. https://doi.org/10.1016/j.socec.2018.03.004

Skagerlund, K., Lind, T., Strömbäck, C., Tinghög, G., & Västfjäll, D. (2018b). Financial literacy and the role of numeracy-How individuals' attitude and affinity with numbers influence financial literacy. *Journal of Behavioral and Experimental Economics, 74*, 18–25. https://doi.org/10.1016/j.socec.2018.03.004

Suyanto, S., Setiawan, D., Rahmawati R., & Winarna J. (2021). Attitudes and behaviour of students towards environmental concerns. *Procedia Environmental Science, Engineering and Management, 8*(4), 1039–1046.

Talwar, M., Talwar, S., Kaur, P., Tripathy, N., & Dhir, A. (2021). Has financial attitude impacted the trading activity of retail investors during the COVID-19 pandemic? *Journal of Retailing and Consumer Services, 58*, 102341. https://doi.org/10.1016/j.jretconser.2020.102341

Van Der Cruijsen, C., De Haan, J., & Roerink, R. (2021). Trust in financial institutions: A survey. *Journal of Economic Survey*. https://doi.org/10.1111/joes.12468

Van Rooij, M., Lusardi, A., & Alessie, R. (2012). Financial literacy, retirement planning and household wealth. *The Economic Journal, 122*, 449–478. https://doi.org/10.1111/j.14680297.2012.02501.x

Wagland, S. P., & Taylor, S. T. (2009). When it comes to financial literacy, is gender really an issue? *The Australasian Accounting Business and Finance Journal, 3*(1), 13–23.

Walstad, W. B., & Soper J. C. (1983). Measuring economic attitudes in high school. *Theory & Research in Social Education, 11*(1), 1983.

Yogasnumurti, R. R., Sadalia, I., & Irawati, N. (2020). The effect of financial attitude and financial knowledge on personal financial management of university students moderated by gender. *International Journal of Research and Review, 7*(2), 219–229.

Yong C. C., Yew S. Y., & Wee C. K. (2018). Financial knowledge, attitude and behaviour of young working adults in Malaysia. *Institutions and Economies, 10*(4), 21–48.

Yurt, E. (2014). The predictive power of self-efficacy sources for mathematics achievement. *Education and Science, 39*(176), 159–169.

Zan, R., & Di Martino, P. (2007). Attitude toward Mathematics: Overcoming the positive/negative dichotomy. In B. Sriraman (Ed.), *The montana mathematics enthusiast* (Monograph 3, pp. 157–168). The Montana Council of Teachers of Mathematics.

CHAPTER 7

Does Financial Literacy Progress Over Time? An Analysis of Three Surveys in Italy

Paola Bongini, Doriana Cucinelli, and Mariangela Zenga

7.1 Introduction

Since the global financial crisis, financial literacy has become an essential life skill in modern economies (OECD, 2020). Indeed, in the last two decades, financial services and products have become more complex but more easily accessible by final users, with the intermediation role of

P. Bongini
Department of Business and Law, University of Milano Bicocca, Milan, Italy
e-mail: paola.bongini@unimib.it

D. Cucinelli (✉)
Department of Economics and Management, University of Parma, Parma, Italy
e-mail: doriana.cucinelli@unipr.it

M. Zenga
Department of Statistics and Quantitative Methods, University of Milano Bicocca, Milan, Italy
e-mail: mariangela.zenga@unimib.it

P. Wachtel et al. (eds.), *Creating Value and Improving Financial Performance*, Palgrave Macmillan Studies in Banking and Financial Institutions, https://doi.org/10.1007/978-3-031-24876-4_7

financial consultants much weakened or even absent. At the same time, financial markets are changing rapidly, with new technologies and the spread of new financial providers in a FinTech ecosystem. Moreover, individuals are increasingly called on to make more financial decisions than before, including planning for retirement or investing in additional education (Lusardi, 2019). For these reasons, financial education today plays a vital role in the policy agenda. It is globally recognized that financial literacy is an essential indicator of an individual's ability to make a wise financial decision and that financially literate individuals are a vital ingredient for the wealth and health of an economy and society as a whole (Lusardi & Mitchell, 2014).

At the national and international level, several surveys exist nowadays that help measure, compare, and detect the determinants of the level of financial literacy of targeted individuals (adult population, teenagers, college students).

However, although studies on financial literacy have multiplied in recent years, to the best of our knowledge, there have been few empirical analyses on the evolution and determinants of the change in the level of financial literacy of a population. In fact, to date, the literature has mainly investigated the differences in financial literacy between countries (Atkinson & Messy, 2012; Borodich et al., 2010; Di Salvatore et al., 2018; Lusardi, 2019; Lusardi & Mitchell, 2014; Nicolini et al., 2013). These studies underline that the level of financial literacy across countries differs quite widely, although sharing a disappointing fact, i.e., being low.

This study aims to fill this gap by investigating the evolution of Italy's level of financial literacy over time and the determinants that explain such change.

Little research has investigated how financial literacy changes over time as noted above. Most literature consists of reports showing the evolution of the level of financial literacy. It tends not to focus on the determinants of changes over time or the causal effects of financial literacy, which is mainly measured in terms of financial knowledge, on financial outcomes such as retirement planning, stock market participation, ability to meet unexpected expenses, wise credit card behavior, etc. For instance, the FINRA Foundation's National Financial Capability Study (NFCS) of 2019 included an annual report on the financial literacy of the US population. Results show an evident decline in financial literacy over nine years, from the first survey in 2009 to the last in 2018, though the differences in successive waves are minor.

Moreover, this study highlights that the decrease in financial literacy is more prominent for younger and middle-aged adults (18–34 and 35–54 years old) and smaller for adults over 55. The authors explain this by the disappearance from public awareness of the high-inflation, high-interest rate environment of the 1980s. Most Americans, except those who experienced that period firsthand as adults, are becoming less aware of these fundamental issues, which are central to the NFCS financial literacy quiz.

Taking stock of the same longitudinal data, Angrisani et al. (2020) investigate the evolution of financial literacy over time and shed light on the causal effect of financial knowledge on financial outcomes. Over a six-year observation period, they find that financial literacy has significant predictive power for future financial outcomes, even after controlling for baseline financial characteristics and a comprehensive set of demographic and individual factors that influence financial decision-making. Schmeiser and Seligman (2013) reach a different result. Using longitudinal data from the US Health and Retirement Study, they examine whether some of the questions previously used as measures of financial literacy are consistent measures of financial knowledge and effective predictors of future changes in wealth. They find that individuals' answers to financial literacy questions show a great degree of inconsistency over time. Good performance in a financial literacy quiz has little predictive power for future accumulated wealth or resilience to financial shocks. Outside the US, PISA, the OECD's Programme for International Student Assessment, an international longitudinal study, has measured the level of financial literacy of teenagers, among other indicators, every three years since 2012. Findings for 2012, 2015, and 2018 (https://www.oecd.org/pisa/) highlight that, on average across OECD countries/economies, mean performance in financial literacy did not change significantly between 2012 and 2018, although it improved by 20 score points between 2015 and 2018. In the case of Italian teenagers, the average level of financial literacy is stable and low, throughout the period, despite significant efforts made by public and private institutions to invest in financial education programs.

Although panel surveys depict a situation where financial literacy is either stable or even declining over time, no study has attempted to define the determinants of these disappointing outcomes. To fill this gap in the literature, this study analyses the evolution of the financial literacy index of Italian adults measured in 2013, 2017, and 2020, disentangling the evolution of its main components (financial attitude, financial behavior, financial knowledge) as described by the OECD,[1] by year and socio-demographic characteristics of the sample. Univariate and Classification and Regression Tree (CART) analyses are used to achieve this.

More specifically, our study addresses the following research questions: (a) Has the level of financial literacy in Italy improved over the last decade?; (b) Does the change in financial literacy correlate with the socioeconomic and socio-characteristics of respondents to identify specific clusters in need of financial education?; and (c) Do these clusters change over time as financial literacy changes?

To answer these questions, we collect information from the three representative surveys on Italian adults' financial literacy and competences conducted in 2013 (PattiChiari and a group of Italian Universities), in 2017, and in 2020 (Banca d'Italia). The data collected makes it possible to measure an overall financial literacy index and highlight its main components (knowledge, behavior, and attitude). Socio-demographic and socioeconomic characteristics of Italian adults are also included and matched with the financial scores.

It is found that despite the proliferation of educational programs, the financial literacy of Italian adults worsened during the period considered. Although the financial knowledge component shows a slight improvement since 2017, the drop in the level of financial attitude, but above all in the level of financial behavior, yields a disappointing evolution of the overall score of financial literacy in Italy. Among the socio-demographic explanatory variables, the worsening of the financial literacy level of younger people is particularly striking. Despite these frustrating outcomes, some improvements are worth mentioning. There is, for example, the closure of the gender gap where women are at a disadvantage on financial issues. First, Italian women's attitude toward money and saving is similar to that of men in all three survey rounds. Second, importantly, a slow and persistent improvement characterizes female financial knowledge, compared to the declining trend among males. The multivariate analysis highlights that explanatory factors change over time and concern: (a) *the survey year*, with 2013 adults being in a better position compared to subsequent surveys; (b) *the level of education*, with highly educated individuals better off in 2013 in terms of financial literacy; (c) *the employment status*, which also hides a generational issue since the least

[1] https://www.oecd.org/daf/fin/financial-education/measuringfinancialliteracy.htm.

financially literate are students and individuals in search of a job, i.e., mainly younger people; and (d) *geographical area*, with adults (and in particular, women) living in the North-Eastern part of Italy being in a better position.

The paper makes several contributions to knowledge in the field. As noted above, to the authors' knowledge, this is the first paper to analyze the evolution of financial literacy of an adult population over time. Secondly, thanks to the availability of survey data over a seven-year horizon, our study of the determinants of financial literacy is extensive and enriched compared to the extant literature. Finally, our pooled cross-sectional study shed light on the evolution of the clusters of the adult population that is in major need of educational programs.

The remainder of the paper is organized as follows. Section 7.2 describes the survey instruments and the sample. Section 7.3 depicts the intertemporal evolution of the financial literacy index and its components. Section 7.4 contains the empirical analysis. Section 7.5 reports the results of our empirical methods, and Sect. 7.6 concludes.

7.2 Survey Instrument and Sample

7.2.1 Survey Instrument

Methods to measure financial literacy vary according to the conceptual definitions used, encompassing different sets of knowledge, skills, and behaviors covering various financial topics. Topics include budgeting, managing money, credit, and debt effectively; assessing the needs for insurance and protection; evaluating the different risks and returns involved in savings and investment options; saving for long-term goals; and understanding the capital market system and financial institutions. Since 2009, the OECD International Network on Financial Education (INFE) has developed a survey instrument to capture people's financial literacy from different backgrounds in a wide range of countries. The survey comprised good practice questions drawn from existing financial literacy questionnaires.[2]

The OECD-INFE has defined financial literacy as "*A combination of awareness, knowledge, skill, attitude, and behavior necessary to make sound financial decisions and ultimately achieve individual financial wellbeing*",

[2] https://www.oecd.org/daf/fin/financial-education/measuringfinancialliteracy.htm.

and the core questions in the survey cover those aspects of knowledge, behavior, and attitudes that are associated with the overall concept. The questions include a range of contexts, including accessing financial services, budgeting, and money management, and planning for the future. There are also questions on important socio-demographic details of the participants, including age, gender, and income. Almost all the questions relate directly to the individual answering the question. However, some information is collected about the household, including total household income and the number of people living with the respondent. Finally, the questionnaire was designed to be used in face-to-face or telephone interviews.

After its release and widespread use in national surveys,[3] the questionnaire was first revised in 2015 (OECD-INFE, 2015) and again in 2018 (OECD-INFE, 2018). These revisions do not, however, prevent comparisons of survey results across years, and precautions are taken relating to this, and are briefly described below. Certain modifications, additions, and deletions were deemed appropriate because the state of knowledge and the financial landscape change rapidly. The questionnaire needs to provide cross-comparable data on emerging and important topics, such as digital financial services, crypto-assets, trust, integrity, and financial consumer protection, while still providing the depth of information necessary to inform a national strategic approach financial education (OECD-INFE, 2018).

Based on the OECD-INFE first version of the questionnaire, at the beginning of 2014 a consortium of Italian banks (PattiChiari) and a group of universities plus a research Centre (Invalsi) ran the first wave of the survey on 1000 adult individuals (http://www.feduf.it/container/scuole/ricerche); the two following waves (2017 and 2020) of the survey were run by the Bank of Italy based on the harmonized (2015 version) and the revised questionnaire (2018 version) on approximately 2500 adult individuals (di Salvatore et al., 2018).

As defined by the OECD, financial literacy is composed of three distinct aspects: knowledge, behavior and attitudes, which are measured by three respective indexes, i.e., a financial knowledge index (FKI); a financial attitude index (FAI), and a financial behavior index (FBI).

[3] By 2012, the Core Questionnaire had been used in 14 countries (Atkinson & Messy, 2012). By 2016, this number increased to 30 countries (OECD, 2016).

The knowledge component aims to assess the understanding of basic concepts that are a prerequisite for making sound financial decisions. Knowledge is based on the topics that have become the standard in the literature on financial literacy (Lusardi & Mitchell, 2014): understanding simple and compound interest, inflation, the positive relationship between financial risk and financial return, and the benefits of portfolio diversification. The FKI ranged from 0 to 8 in the first version of the toolkit and from 0 to 7 in the most recent; it is calculated as the sum of correct answers to the set of financial knowledge questions where each correct answer counts as one point and each wrong answer counts as zero. In the survey used in the 2013 wave, the research group eliminated questions that were merely testing a simple mathematical skill (e.g., division) or testing the same concept with two different instruments. The 2013 questionnaire contained only five questions related explicitly to calculating simple and compounding interest (two questions), understanding how inflation works (one question), the link between risk and return (one question), and the power of risk diversification (one question). In order to guarantee the comparability across waves, the FKI is constructed considering only these five questions (see Annex 1 for details).

The second component measures how a person's behavior impacts on his/her financial well-being. Greater ability to properly manage financial resources reflects higher financial literacy. In particular, the behavior index is based on questions assessing whether people manage family financial resources by planning a budget, are able to make ends meet while paying debts and utilities with no concerns, and acquire information before making investments. As above, the financial behavior index counts positive behaviors exhibited; it takes a maximum value of 9, and a score of 6 or more is considered to be relatively high. Of the three indexes, the FBI is the one that has undergone the biggest changes in the formulation of questions and subsequent construction of the index itself (see Annex 1 for details) also for the two subsequent waves (2017 and 2020). As a consequence, in the descriptive statistics, we propose two different versions of the index so as to guarantee its comparability through the years.

Finally, the attitudes component evaluates preferences, beliefs and non-cognitive skills which are likely to affect personal well-being. Following INFE methodology, this component is meant to capture attitudes toward precautionary saving and planning for the future. If individuals have a negative attitude toward saving for their future, for example, it is argued that they will be less inclined to save. Similarly, if they prefer to prioritize short-term wants, they are unlikely to save for an emergency or make long-term financial plans (Atkinson & Messy, 2012). Therefore, the financial literacy survey includes three scaled attitudinal questions that ask people about whether they agree or disagree (on a scale from 1 to 5) with particular statements that capture their disposition or preferences: "I find it more satisfying to spend money than to save it for the long term", "I tend to live for today and let tomorrow take care of itself", and "Money is there to be spent". The FAI is created by adding together the responses to each of the three questions, and then dividing by 3, so the score ranges between 1 and 5. The score is considered to be high when it is above 3. In general, individuals who disagree with the statements tend to have a longer-term view.

In order to assess overall levels of financial literacy, the three indexes are added, giving a single measure that considers the various aspects of financial literacy, including financial planning for the future, choosing financial products, and managing money on a day-to-day basis. The Financial Literacy Score can take a minimum value of 1 and a maximum value of 19, given by a maximum of 5 points from the knowledge index, 9 from behavior, and 5 from attitudes. There are no penalties for wrong answers, so "I don't know" and no response are treated as "wrong" answers.

Waves 2 and 3 of the Italian survey also include questions to assess the respondents' level of self-confidence. These were not present in the first version of the questionnaire and will be analyzed below in discussing responses and differences between 2017 and 2020.

7.2.2 *Sample*

As detailed in the OECD-INFE toolkit, (i) the survey should be of adults, i.e., individuals aged between 18 and 79; (ii) the interviews should preferably be made by telephone or face to face, to overcome issues related to low levels of literacy; and (iii) a minimum sample size of 1000 participants per country should be collected for international comparisons.

The three rounds of the Italian survey followed these requirements, and the samples were stratified per quota based on gender, age, geographical area, and municipality size. The samples were thus representative of the Italian population with regards to gender, age, geographical location, and the dimension of municipalities. Table 7.1 reports the distribution of the samples relative to main socio-demographic variables.

Table 7.1 Sample distribution

		2013		*2017*		*2020*	
	Categories	*Frequency*	*%*	*Frequency*	*%*	*Frequency*	*%*
Gender	Male	598	47.96	1140	47.98	1056	51.9
	Female	649	52.04	1236	52.02	980	48.1
Age	18–24	112	8.98	240	10.10	160	7.9
	25–34	187	15.00	310	13.04	271	13.3
	35–44	237	19.01	419	17.63	326	16.0
	45–54	218	17.48	466	19.60	392	19.3
	55–64	187	15.00	325	13.67	332	16.3
	65 and over	306	24.54	617	25.96	554	27.2
Geographical Area	North-West	343	27.51	634	26.68	544	26.7
	North-East	237	19.01	456	19.19	393	19.2
	Central Italy	249	19.97	476	20.03	407	20.0
	Southern Italy and Islands	418	33.52	810	34.09	691	33.9
Educational attainment	Primary school or lower	274	21.97	260	10.94	246	12.1
	Lower secondary school	449	36.01	916	38.55	520	25.5
	Upper secondary school	387	31.03	739	31.10	953	46.8
	High school or above	137	10.99	461	19.40	317	15.6
Job status	Employed	525	42.13	1054	44.38	1019	50.1
	Unemployed seeking employment	103	8.27	241	10.15	126	6.2
	Inactive not seeking employment	618	49.60	1080	45.47	889	43.7
Data collection	CATI	1247	100	–	–	–	–
	TABLET	–	–	1178	49.58	–	–
	CAPI	–	–	1198	50.42	2034	100
	Total	1247	100	2376	100	2034	100

Table reports frequency values of the surveyed sample with respect to main socio-demographic characteristics and methods of data collection

In 2013, the sample of 1247 respondents was collected using CATI (Computer Assisted Telephonic Interviews). Of the subjects interviewed, 300 were interviewed by cell phone, and the others by landline. Of the individuals surveyed, 52.04% were female. Most respondents (76.5%) were under 64 years old. The respondents' average educational level was: 31% were educated up to higher secondary level, 11% were university graduates, and 58% had at most a middle-school-level education. Approximately half of the respondents (49.6%) were inactive and not seeking employment; the remaining were unemployed (8.3%) or employed (41.1%).

In 2017, the survey was run on approximately 2376 adults. The survey was carried out using two different methodologies: 49.6% of individuals responded on a tablet designed to be easily used by all population subgroups, including the less educated and the elderly. The others were interviewed personally using CAPI (Computer Assisted Personal Interviews). Compared to the 2013 sample, older adults increased from 25 to 26%. The level of education also increased slightly, with "high school or above" rising from 11 to 19.4%. Approximately 46% of the sample were inactive and not seeking employment; the remaining respondents were either unemployed (10.2%) or employed (44.4%).

The last survey wave, carried out in early 2020, involved an overall sample of about 2000 individuals interviewed using CAPI. The aging trend of the Italian population is confirmed by the constant increase in the percentage of over 65s on the total number of respondents. There was a noticeable drop in the highest level of educational attainment; only 15% obtained a high school diploma or a college degree compared to 19% in the 2017 sample. The percentage of unemployed, either searching for a job or not seeking employment, also fell. There is also a drop in the rate of women interviewed; it fell from 52% in 2013 and 2017 to 48% in 2020.

7.3 The Intertemporal Evolution of Financial Literacy Levels: Descriptive Statistics

The average level of financial literacy of Italians shows a statistically significant negative trend, falling from almost 12% in 2013 to a mere 10% in 2020 (Table 7.2).

The drop in the overall index is attributable to the fall in all its three components, and in particular to FBI. Financial knowledge recorded a small drop of 0.04 points, while behavior and attitude drop by 1.42 and 0.30 points, respectively (Fig. 7.1).

Table 7.3 reports the median, mean values, and standard deviation of the Financial Knowledge Index (FKI) across the three waves. In the period under investigation, the mean value of the FKI decreased from 2.59 points in 2013 to 2.36 in 2017 to return in 2020 close to the initial value of the first survey 2.55. Note that the median value, on the other hand, shows an increase in 2020, from 2 to 3 points, indicating that half of the sample's knowledge has in fact improved. As noted in

Table 7.2 Global Financial Literacy Index

Year interview	*Mean value*	*Sample size*
2013	11.77***	1213
2017	10.35***	2215
2020	10.16***	1894

Table reports the mean value of the Global Financial Index in 2013, 2017, and 2020 and the sample size in the surveys. Anova test for differences in means. Significance ***: <0.0001

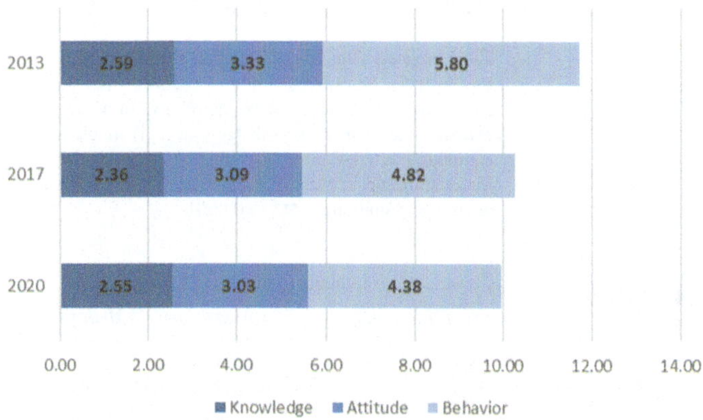

Fig. 7.1 Global Financial Literacy Index evolution (Figure reports the Global Financial Index evolution across 2013, 2017, and 2020. The indicator is made up of three components—knowledge, attitude, and behavior. The differences in means of each indicator across years are always statistically significant at 99%)

Sect. 7.2, the index measures the level of knowledge of basic financial concepts which are considered as a prerequisite for making sound financial decisions. In fact, responding correctly to all five questions does not imply that individuals are financial experts, but only that they know basic concepts of personal finance. The results suggest that on average Italians have knowledge of fewer than half the financial concepts investigated.

There is a significant variation of attitude across surveys. From 2013 to 2020, the Financial Attitude Index decreases from 3.32 to 3.03, suggesting that individuals' attitudes tend less toward the long term (Table 7.4).

Similarly, the FBI shows a decrease of the mean and median values and also in the measure of variability (Table 7.5). As noted in Sect. 7.2, the wording of the items chosen to measure financial behavior changed across the surveys. Specifically, in the 2017 and 2020 surveys, the responses on "active saving" did not include "cash deposited in a bank account" as an option. Differences also emerged with regard to the question about

Table 7.3 Financial knowledge index across the surveys

Year interview	*Mean*	*Median*	*Std. Dev.*
2013	2.59***	2.00	1.31
2017	2.36***	2.00	1.43
2020	2.55***	3.00	1.52

Table reports the mean, median, and standard deviation of the financial knowledge index, assessing the understanding of basic financial concepts. Five questions on basic financial topics compose the financial knowledge index: (i) simple interest rate; (ii) compound interest rate; (iii) inflation; (iv) diversification; and (v) the relationship between risk and return. Anova test for differences in means. Significance ***: <0.0001

Table 7.4 Financial Attitude Index

Year interview	*Mean*	*Median*	*Std. Dev.*
2013	3.32***	3.33	0.91
2017	3.09***	3.00	0.88
2020	3.03***	3.00	0.80

Table eports the mean, median, and standard deviation of the Financial Attitude Index Mean, assessing the attitude toward savings for the future. Anova test for differences in means. Significance ***: <0.0001

Table 7.5 Financial behavior index

Year interview	*Mean*	*Median*	*Std. Dev.*
2013	5.80***	6.00	1.460
2017	4.82***	5.00	1.963
2020	4.38***	5.00	1.825

Table reports the mean, median, and standard deviation of the Financial Behavior, assessing those actions and behaviors of consumers that could determine their financial conditions and well-being in the short and medium to long term. Anova test for differences in means. Significance ***: <0.0001

Table 7.6 Financial behavior index recalculated with no shopping around and active saving items

Year interview	*Mean*	*Median*	*Std. Dev.*
2013	4.77***	5.00	1.06
2017	4.03***	4.00	1.33
2020	3.71***	4.00	1.43

Anova test for differences in means. Significance ***: <0.0001

the ways in which individuals buy their financial products ("shopping around"), by simply relying on the advice of their bank or friends, or making comparisons using independent advice. Table 7.6 thus reports the FBI without the two items, "active saving" and "shopping around", which changed. The drop in the index is less pronounced, with a decrease of 1.06 points. However, the downward trend is confirmed.

The interested reader can refer to Annex 2 for a discussion of the changes item by item in the three components of the Financial Literacy Index

7.3.1 *Socio-demographic Characteristics*

Figures 7.2, 7.3, 7.4, and 7.5 show the evolution of the index of Financial Literacy according to main socio-demographic characteristics of the respondents, education, age, and gender.

Results are in line with international evidence of the determinants of financial literacy levels (Lusardi & Mitchell, 2014). Graduates have a higher degree of financial literacy than individuals with lower educational

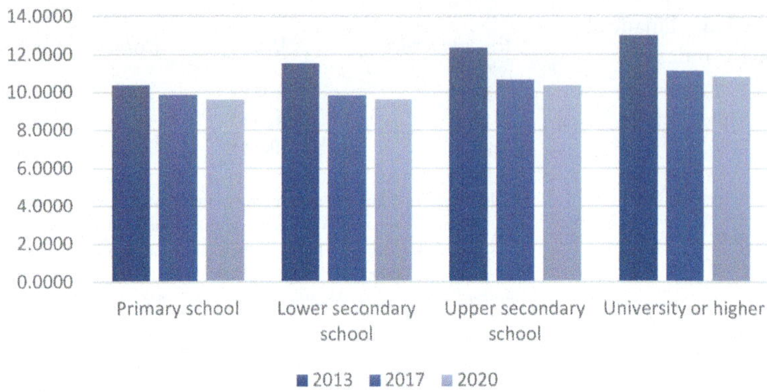

Fig. 7.2 FLI and educational attainment (Figure shows the level of financial literacy index among the different level of education observed in the different surveys)

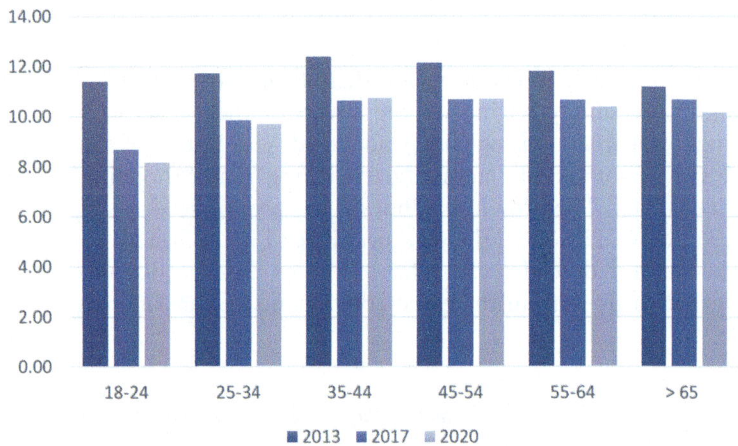

Fig. 7.3 FLI and age distribution (Figure shows the level of financial literacy index among the different age observed in the different surveys)

levels. Financial literacy increases with age up to a turning point, corresponding to the retirement age (Mazzonna & Peracchi, 2020; OECD, 2016) and males are more financially literate than females.

Fig. 7.4 FLI and gender (Figure shows the level of financial literacy index of male and female observed in the different surveys)

It is worth focusing on the temporal evolution with respect to age and gender. The largest drop in the level of financial literacy is recorded for the age bracket "18–24 years", which also presents the lowest level of overall literacy. The FLI has remained stable in recent years for the three age brackets that include individuals at their initial and middle stages of working life.

The gender gap that characterized the population in 2013 seems to have closed in 2017 and 2020, reflecting mainly a marked worsening in male financial literacy. This is clear in the changes in the three components of financial literacy between 2013 and 2020 (Fig. 7.5). Women's attitude toward money and saving is similar to that of men, and this is constant in all three waves. On the other hand, women score lower than men when it comes to demonstrating financial knowledge or savvy behavior, a result found in much literature (Lusardi & Mitchell, 2014). It is, however, worth pointing out the slow and persistent pace of improvement that characterizes female answers to the FK items, as opposed to the decrease for males.

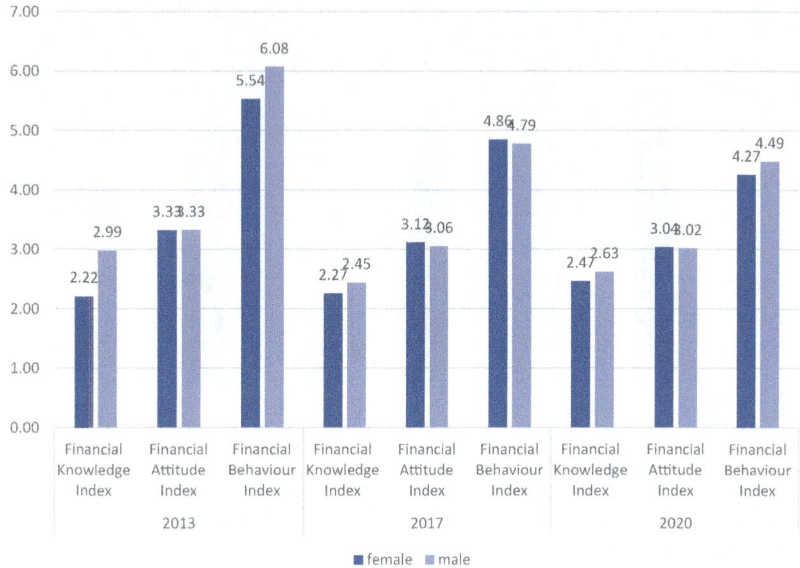

Fig. 7.5 Gender and the components of financial literacy (Figure shows the level of financial knowledge, attitude, and behavior distinguishing between male and female observed in the different surveys)

7.4 Empirical Methods

The descriptive statistics have highlighted a worsening in financial literacy, with the two components of attitude and behavior being the driving forces of such deterioration. This section analyzes the determinants of this evolution over the years through the Classification and Regression Tree methodology that helps define groups of individuals connected by the same level of financial literacy (and similar financial education needs).

7.4.1 CART Analysis

The CART (Classification and Regression Tree) methodology is a non-parametric tree-structured recursive partitioning method introduced by Breiman et al. (1984). In general, it is an alternative approach to nonlinear regression. It is a method that belongs to binary decision tree approaches built by repeatedly splitting a node into two child nodes, beginning with the root node containing the whole sample. The use of CART analysis in this work helps to facilitate the use of covariates where we can explore the influence of many variables on the respondents' variable (Financial literacy index).

In regression tree terms, let T represent the dependent variable and $X = (X_1, X_2, ..., X_p)$ be a vector of p covariates. The method involves two main stages to build the regression tree: growing and pruning. In growing, the T is recursively partitioned into subsets. Each partition is obtained by examining every possible binary split along the observed data T for each predictor variable $X_1, X_2, ..., X_p$ and selecting that which most reduces the variability of the node regarding the predicted variable. The result is a sequence of nested trees, with increasing leaves (terminal nodes), until no more splits are possible and the fully grown tree is reached. The pruning stage of the fully grown tree aims then to select the best sub-tree and consists of declaring an internal node as terminal and deleting all its descendants.

In the analysis, we obtained the tree using the following criteria:

- Minimum number of cases in the parent node: 100;
- Stopping rule for a terminal node: 50;
- Tree pruning to avoid overfitting with a maximum acceptable difference in risk between the pruned and the sub-tree of 3 standard error; and
- Missing data handled by surrogate splits.

The dependent variable of our analysis is the financial literacy index (FLI).[4] Explanatory variables are included and explained in Table 7.7.

7.4.2 Explanatory Variables

In X, we consider socio-demographic, socioeconomic variables, and a variable on the survey's year. As suggested by previous literature (Cucinelli et al., 2019; Lusardi, 2019; Lusardi et al., 2010), among socio-demographic and socioeconomic variables we include: (i) gender, measured with a categorical variable with two categories: female and male;

[4] Annex 3 reports the results of CART analysis applied to the financial knowledge index, the financial attitude index, and the financial behaviour index.

Table 7.7 Variables description

Variables	*Acronym*	*Description*	*Values*
Dependent variable			
Financial literacy index	FLI	Numerical variable on the financial literacy score defined following OECD directions	[2.67; 18.67]
Independent variable			
Respondent's Gender	Gender	Categorical variable on the respondent's gender	1 = Female 2 = Male
Respondent's Age	AGE	Ordinal variable on the respondent's age	1 = [18; 24] years old 2 = [25; 44] years old 3 = [45; 64] years old 4 = $\geq$ 65 years old
Respondent's employment status	Employment condition	Categorical variable on the respondent's employment status	1 = Employee 2 = Self-employed 3 = Student 4 = Retired 5 = Housewife 6 = Person seeking for a job
Respondent's Education	Education	Ordinal variable on the respondent's education level	1 = Primary education 2 = Lower secondary education 3 = Upper secondary education 4 = University or more
Respondent's geographical area	Geographical area (3)	Categorical variable on the Italian geographical area in which the respondent lives	1 = North Italy 2 = Centre Italy 3 = South Italy or Islands

(continued)

Table 7.7 (continued)

Variables	*Acronym*	*Description*	*Values*
Respondent's citizenship	Citizenship	Categorical variable on the respondent's citizenship	1 = Italian 2 = Not Italian
Respondent's family component number	Family component number	Ordinal variable on the respondent's family component number (including the respondent)	1 = One component 2 = Two components 3 = Three components 4 = Four components or more
Survey year	Survey year	Ordinal variable on the year in which the survey was administered	1 = 2013 2 = 2017 3 = 2020

(ii) age, measured using an ordinal variable with four categories: 18–24, 25–44, 45–64, and = ≥65 years old; (iii) citizenship, measured with a categorical variable with two categories: Italian and not Italian; (iv) geographical area in which the respondent lives; (v) family components number, measured with an ordinal variable with four categories: one component, two components, three components, four components, or more; (vi) employment status, measured using a categorical variable with six categories: employee, self-employed, student, retired, housewife, and looking for a job; and (vii) educational level, measured using an ordinal variable with four categories: primary education, lower secondary education, upper secondary education, and university education or more. Moreover, we include an ordinal variable for the survey year with three categories (2013, 2017, and 2020).

Table reports the description of socio-demographic and socioeconomic variables used in the analysis.

7.5 Results and Discussion

Figure 7.6 reports the most significant explanatory variables of the level of financial literacy in the three surveys, in order of statistical relevance. The *Survey year* shows the highest level of importance, followed by *Employment condition, Level of education, Age, Gender,* and *Geographical area.* The *Citizenship* and *number of family components* are the variables with the lowest level of importance.

The final nodes of the FLI pruned tree identify 6 clusters as reported in Table 7.8 and Fig. 7.7 while Table 7.12 in the Annex explores the sociological characteristics of the identified groups. The first cluster includes individuals scoring the highest financial literacy, with an average FLI of 12.459. It consists of 567 respondents of the 2013 survey, with a medium to a high level of educational attainment, essentially young men workers. The second group presents an average FLI equal to 11.011 and is composed of 589 respondents of the 2013 survey with a medium to low level of education, mainly non-working older women. These two nodes underline that in 2013 the level of education was the first discriminant that described the level of financial literacy. Looking at the two other surveys (2017 and 2020), our results suggest that the level of education becomes less critical, leaving room for the type of employment of respondents. This may indicate that the many financial education

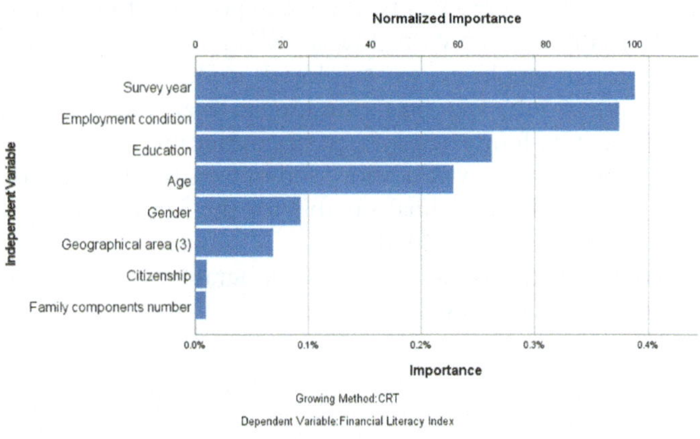

Fig. 7.6 Main explanatory factors of the Financial Literacy tree (Figure shows the most significant explanatory factors of the financial literacy index)

programs carried out since 2013 have reduced the differences among individuals with different levels of education. These programs were mainly addressed to students of primary and secondary schools and women in general, increasing the level of financial literacy of these two parts of the population.

In the last two surveys, the employment status becomes the most critical discriminant. Retired, self-employed, employed, and housewives show a higher level of financial literacy than students and people looking for a job. This classification also hides an age difference between the two clusters; cluster 3 comprises young people, aged 18–24 years, equally distributed between males and females. The level of education returns to be important when we look at the differences within the group of employed, housewives, retired, and self-employed individuals, which comprises so many different types of individuals by their working status. Again, individuals with a higher educational level show a higher financial literacy, 10.924. Finally, for those less literate, the geographical area where they live seems essential in describing the differences in financial literacy. In general, respondents who live in the South of Italy show a lower financial literacy index (9.358). In comparison, individuals who live in the North and center of Italy show a higher financial literacy index (10.290). Interestingly, the three subsequent clusters identified by the CART are in the majority composed of women, where the first discriminating factor is being in the job market (cluster 4) or not (cluster 5 and 6) and subsequently the residence area (North versus South), confirming

Table 7.8 Description of the six final groups by the regression tree on FLI

Cluster	*n*	*Std. Dev.*	*Mean*	*Description*
Cluster 1	589	2.336	11.011	Year: 2013; Education ≤ Secondary school
Cluster 2	567	2.389	12.459	Year: 2013; Education > Secondary school
Cluster 3	643	2.669	8.833	Year: 2017/2020; Students, Person looking for a job, Other work condition
Cluster 4	1919	2.925	10.924	Year: 2017/2020; Education > Secondary school
Cluster 5	1056	2.454	10.29	Year: 2017/2020; Education: ≤ Secondary school; North Centre
Cluster 6	543	2.518	9.358	Year: 2017/2020; Education: ≤ Secondary school; North Centre

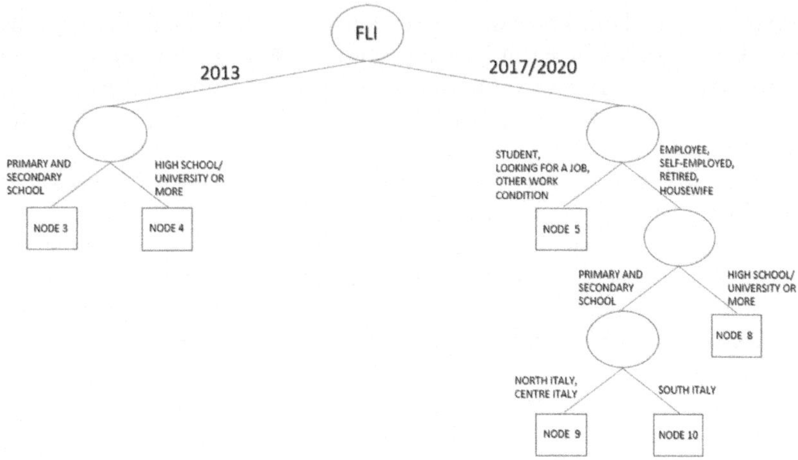

Fig. 7.7 Tree diagram of Financial literacy indexes–in figure "2020" instead of "2022" (Figure reports the stylized CART with the six clusters)

the results of Cucinelli et al. (2019) who highlighted the importance of local factors in shaping the financial literacy level of Italian adults.

Table reports the characteristics of the six clusters defined by the regression tree.

To summarize, the CART results suggest significant differences between groups of individuals. Differences emerge concerning: (a) *the survey year*, with 2013 adults being in a better position compared to subsequent surveys; (b) *the level of education*, with highly educated individuals better off in terms of financial literacy; (c) *employment status*, which also hides a generational issue, since the least financially literate are students and individuals in search of a job, i.e., mainly younger people; and (d) *geographical area*, with adults (and in particular, women) living in the North-Eastern part of Italy being in a better position.

Our results further confirm that the "*one-size-fits-all*" strategy fails in education. In reality, financial education programs need to be planned and designed, considering the differences mentioned above. It is crucial to tailor educational programs to specific audiences with similar characteristics. In particular, in the most recent surveys, it emerges that the most fragile in terms of financial literacy are those with lower educational attainment, whatever their occupational status (employed or self-employed versus housewife or retired).

7.6 Conclusions

Previous literature has focused on the determinants of financial literacy, considering both socio-demographic and socioeconomic characteristics of individuals. A more recent strand of literature has analyzed the macro-ecological variables that characterize the context in which individuals live. What these studies have in common is that they focus on survey data from just one year. Our contribution to the literature is to study the evolution of financial literacy and its components over a more extended period, at three points in time (2013, 2017, and 2020).

To our first research question (Has the level of financial literacy in Italy improved over the last decade?), we provide evidence of a negative answer. However, among disappointing results, the reduction of the gender gap, thanks to an improvement in female financial knowledge, can be considered a glimmer of light.

About our second and third research questions (Does financial literacy measured over time correlate with the socioeconomic and socio-characteristics of respondents?; Do these clusters change over time as financial literacy changes?), our results underline that these factors are indeed useful in identifying clusters homogeneous in their need of financial education. And more importantly, the clusters change over time as financial literacy changes: if in 2013 the most discriminant variable was the level of education, in the subsequent surveys (2017 and 2020), the level of education gives way to the employment condition, suggesting that the programs of financial education carried out since the first survey were able to reduce the differences among people with different level of education. In more recent years, the employment condition becomes more important in discriminating among Italian adults, with students and individuals looking for a job being less financially literate than others. This result provides important support to the very recent attempts of universities and the national strategy for financial education to target college students of noneconomic fields with personal finance courses. Finally, we provide further support to those studies which underlined the role and impact of local factors in defining the financial literacy of respondents (Cucinelli et al., 2019; De Beckker et al., 2020).

These findings are important to define future financial education programs. As well known, the "One-size-fits-all" programs cannot be successful considering socio-demographic differences highlighted by our analysis. In future, considering the differences in terms of employment status and geographical area in which individuals live becomes crucial to decrease the differences in the financial literacy levels. Moreover, our results underline that the importance of the socioeconomic explanatory factors also changes over time; repeated baseline surveys of financial literacy of the adult population are therefore crucial for designing effective financial education programs.

Given the disappointing results of our research, it is crucial today to rethink and reflect on the structure and content of educational initiatives to improve their performance and achieve their final goal of promoting the financial literacy of Italian adults.

Along this line, the discussion should focus on how to structure new financial education initiatives, whether they are designed as life-long programs or as simply on the job initiatives; whether it is more useful to lever digital and user-friendly modules or traditional face-to-face lectures; and finally, whether non-cognitive approaches are needed (as in Bocchialini et al., 2022, this volume). Most importantly, initiatives should be planned, designed, and monitored according to the best practices outlined for the implementation of effective educational programs. Two relevant references in this regard are the guidelines proposed by the Italian National Strategy for Financial Education (*Comitato per la programmazione e il coordinamento delle attività di educazione finanziaria*) and the fifteen indicators developed by a multidisciplinary team of scholars for the National Observatory of Economic and Financial Education (ONEEF)[5] as useful instruments to design—ex-ante—and evaluate—ex-post—the effective financial education projects.[6]

Future research should focus on the evolution of financial literacy considering panel data on a sample of the same individuals that in

[5] ONEEF is a National Observatory of Economic and Financial education. It was founded in 2016 by an inter-university pool of scholars and practitioners with different disciplinary background (economists, sociologist, pedagogists, and psychologist.). It has three main goals: (a) to monitor with a standardize procedure all the project on financial and economic education run in Italy and provide public data for free to those who are interested in the field; (b) to provide guidelines to improve the quality of the design of financial education projects; and (c) to sustain networking among public, private, and ONG subjects which run economic and financial education projects in Italy and, if possible, abroad.

[6] Each area has detailed questions to help designing an effective project (for example, "*Are the goals following the S.M.A.R: T. model?*", "*Do you know the level of recipients' financial literacy?*" "*what kind of monitoring or evaluation come along with the project?*" https://oneef.unimib.it/i-15-indicatori-oneef/.

different years attend the survey, also asking them whether they have attended educational programs during the period between the two surveys. Moreover, when comprehensive databases on financial education programs become available, future research can use more precise measures for such programs.

Annex 1

The construction of the financial literacy index and its components

The overall Financial Literacy Index is the sum of points scored in each of the three components of the index itself: financial knowledge, financial attitude and financial behavior.

Financial knowledge ranges from 0 to 5 points

Text	*Possible responses*	*Purpose*
Imagine that 5 brothers have to wait for one year to get their share of $1000 and inflation stays at 1%. In one year's time will they be able to buy...	Multiple choice [correct response "less than they could buy today"—1 point]	To test the ability to understand how inflation impacts on purchasing power
Suppose you put $100 into a savings account with a guaranteed interest rate of 2% per year. You don't make any further payments into this account and you don't withdraw any money. How much would be in the account at the end of the first year, once the interest payment is made?	Open response [correct response $102—1 point]	To test the ability to calculate simple interest on savings
And how much would be in the account at the end of five years [add if necessary: remembering there are no fees or tax deductions]? Would it be...] This question builds on previous question	Multiple choice [correct response more than $110—1 point]	To test whether the respondent is aware of the additional benefit of compounding

(continued)

(continued)

Text	*Possible responses*	*Purpose*
An investment with a high return is likely to be high risk/or If someone offers you the chance to make a lot of money it is likely that there is also a chance that you will lose a lot of money	True/False [correct response is true—1 point]	To test whether the respondent understands the typical relationship between risk and return
It is usually possible to reduce the risk of investing in the stock market by buying a wide range of stocks and shares/or It is less likely that you will lose all of your money if you save it in more than one place	True/False [correct response is true—1 point]	To test whether the respondent is aware of the benefit of diversification

Financial attitude score ranges from 1 to 5 (sum of the points scored in each item and then divided by 3)

Text	*Possible responses*	*Purpose*
I find it more satisfying to spend money than to save it for the long term I tend to live for today and let tomorrow take care of itself' Money is there to be spent	5 point scale: 1 = completely agree; 5 = completely disagree	These questions are intended to indicate whether the respondent focuses exclusively on the short term (agrees) or has a preference for longer-term security (disagrees)

Financial behavior ranges from 0 to 9
(sum of 8 items each of them 0 or 1 point with the exception of question "Choosing products" that takes the values 0, 1 or 2)

Text	*Possible responses*	*Value toward final score*
Considered purchase	5 point scale: 1 = completely agree; 5 = completely disagree	1 point for respondents who put themselves at 1 or 2 on the scale. 0 in all other cases
Timely bill payment	5 point scale: 1 = completely agree; 5 = completely disagree	1 point for respondents who put themselves at 1 or 2 on the scale. 0 in all other cases
Keeping watch of financial affairs	5 point scale: 1 = completely agree; 5 = completely disagree	1 point for respondents who put themselves at 1 or 2 on the scale. 0 in all other cases
Long term financial goal setting	5 point scale: 1 = completely agree; 5 = completely disagree	1 point for respondents who put themselves at 1 or 2 on the scale. 0 in all other cases
Responsible and has a household budget	*YES/NO*	1 point if personally or jointly responsible for money management and has a budget. 0 in all other cases
Borrowing to make ends meet	This is a derived variable that combines a question about running short of money and one that identifies a range of different ways in which the respondent made ends meet the last time they ran short of money. The derived variable indicates people who are making ends meet without borrowing	0 if the respondent used credit to make ends meet. 1 in all other cases
Active saving	This question identifies a range of different ways in which the respondent may save. People who refused to answer score 0	1 point for any type of active saving (excluding letting money build up in a current account as this is not active). 0 in all other cases

(continued)

(continued)

Text	*Possible responses*	*Value toward final score*
Choosing products	This is a derived variable drawing information from 2 questions. It is only possible to score points on this measure if the respondent had chosen a product: those with no score on this measure have either refused to answer, not chosen a product, or not made any attempt to make an informed decision	1 point for people who had tried to shop around or gather any information. 2 points for those who had shopped around and gathered independent information. 0 in all other cases

Annex 2

Evolution of FKI, FBI and FAI item by item
Financial Knowledge Index

	2013		*2017*		*2020*	
Questions	*Correct (%)*	*DNK*[7] *(%)*	*Correct (%)*	*DNK (%)*	*Correct (%)*	*DNK (%)*
The effects of inflation	63.00	18.0	47.80	19.3	50.47	16.4
Simple interest rate calculation	33.18	34.3	46.37	31.1	59.45	25.3
Compound interest rate calculation	35.00	23.4	32.20	23.6	28.99	21.1
The power of diversification	46.52	23.3	36.35	38.3	51.31	24.5
The relationship between risk and return	81.70	6.7	73.35	17.3	64.74	17.2

Note Table reports the Financial Knowledge Index as the um of correct answers—index spans from 0 to 5—and the percentage of correct answers given by interviewed people.

[7] The Don't Know answer includes the option "I prefer not to answer".

The Financial Knowledge Index declines due to the reduction of the individuals' knowledge of all topics composing the index, except for the simple interest rate item.

Financial attitude index

The FAI is measured using three different sentences that evaluate the attitude towards saving individuals (money, planning, and future). Results are reported in Figures 7.8, 7.9, and 7.10.

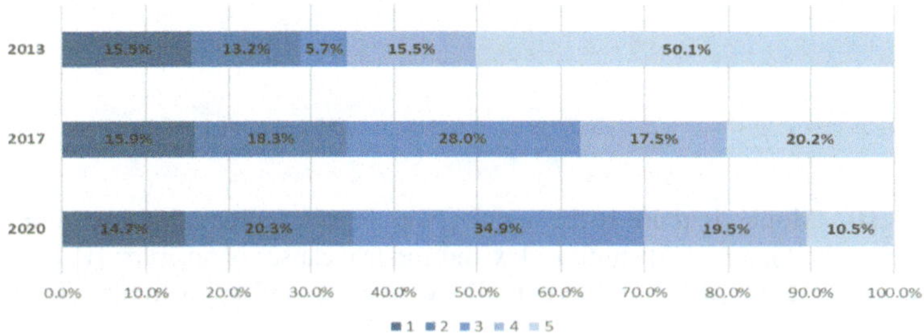

Fig. 7.8 I tend to live for today and let tomorrow take care of itself (*Note* Figure reports the first item that composes the Financial Attitude Index and the five classes of answers: [i] 1 totally agree; [ii] 2 agree; [iii] 3 indifferent; [iv] 4 disagree; and [v] 5 totally disagree)

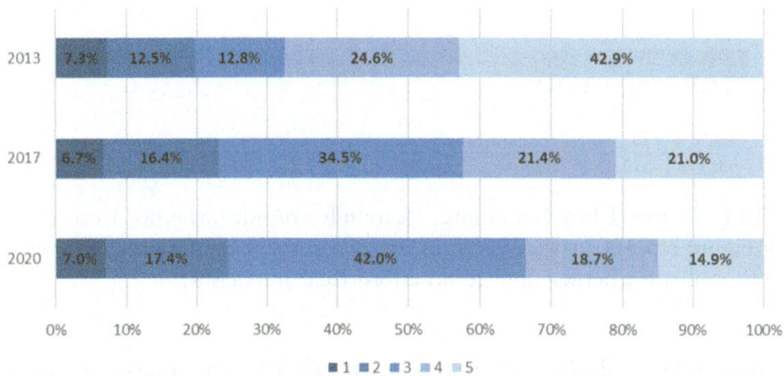

Fig. 7.9 I find it more satisfying to spend money than save it for the long term (*Note* Figure reports the first item that composes the Financial Attitude Index and the five classes of answers: [i] 1 totally agree; [ii] 2 agree; [iii] 3 indifferent; [iv] 4 disagree; and [v] 5 totally disagree)

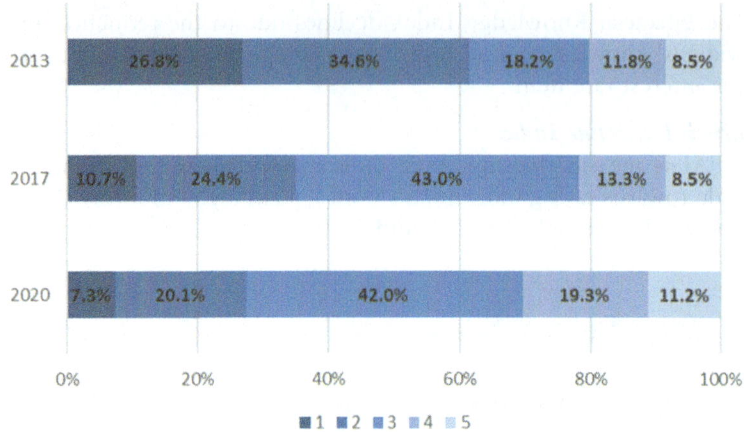

Fig. 7.10 Money is there to be spent (*Note* Figure reports the first item that composes the Financial Attitude Index and the five classes of answers: [i] 1 totally agree; [ii] 2 agree; [iii] 3 indifferent; [iv] 4 disagree; and [v] 5 totally disagree)

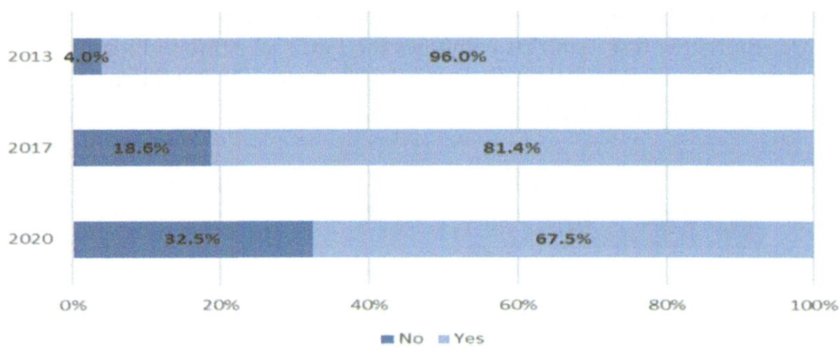

Fig. 7.11 Before I buy something, I carefully consider whether I can afford it (*Note* Figure reports answers given in the two surveys by respondents that were asked to evaluate whether they could afford their purchases)

From 2013 to 2020, the financial attitude of individuals decreases, in terms of both attitude towards savings and consideration of the future.

Financial Behavior Index

See Figures 7.11, 7.12, 7.13, 7.14, 7.15, 7.16, 7.17, and 7.18

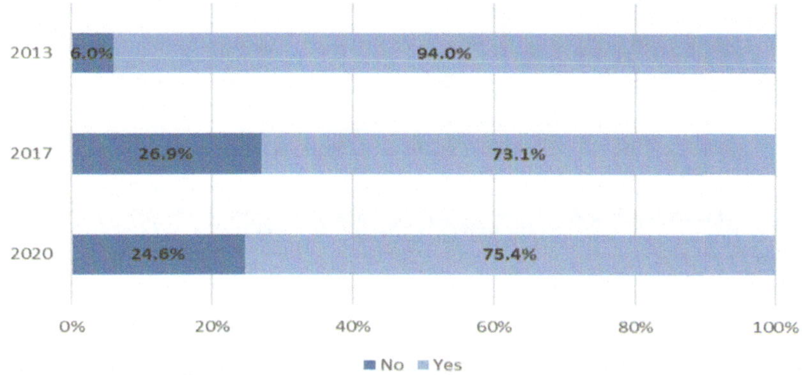

Fig. 7.12 I pay my bills on time (*Note* Figure reports answers, where one is given in the two surveys by respondents that were asked to answer if they pay their bills on time)

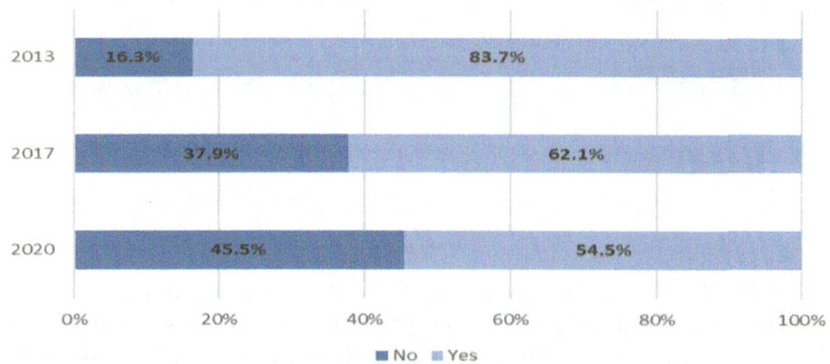

Fig. 7.13 I keep a close personal watch on my financial affairs (*Note* Figure reports answers given in the two surveys by respondents that were asked to evaluate if they keep a close personal watch on their financial affairs)

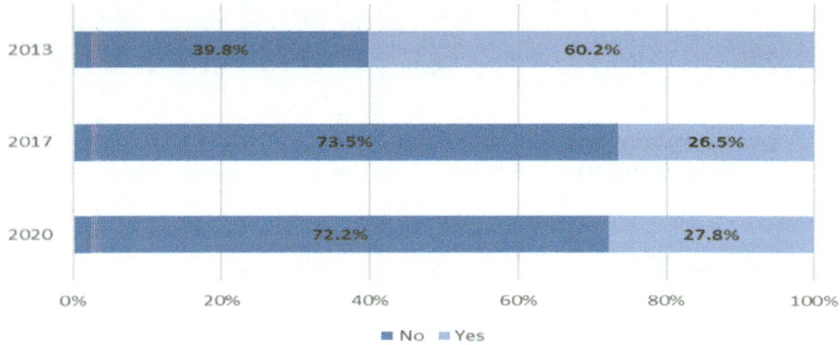

Fig. 7.14 I set long-term financial goals and strive to achieve them (*Note* Figure reports answers given in the two surveys by respondents asking if they set long-term financial goals and strive to achieve them)

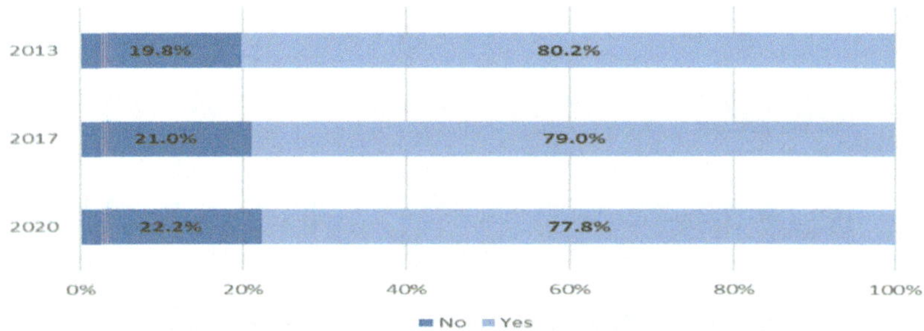

Fig. 7.15 I am responsible for making day-to-day decisions about money in my household (*Note* Figure reports answers given in the two surveys by respondents that were asked to evaluate if they are responsible for making day-to-day decisions about money in their household)

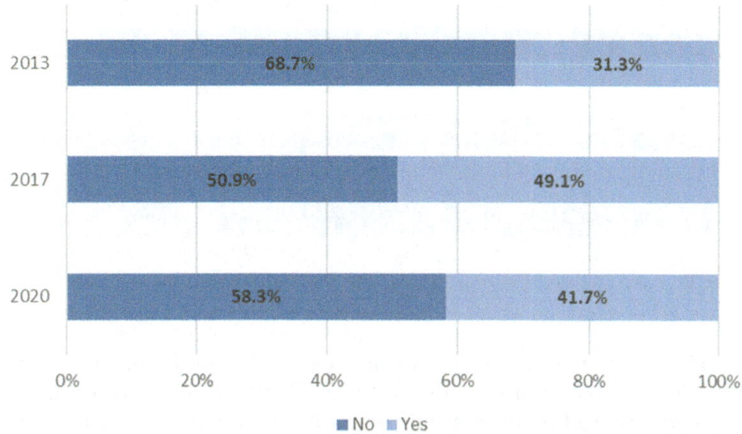

Fig. 7.16 Active saving (*Note* Figure reports answers given in the two surveys by respondents asking if they have some form of active saving. In 2013 surveys, the item provided the following answers: saving cash at home or in your wallet; building up a balance of money in your bank current account; paying money into a savings account; buying financial investment products, other than pension funds; or in some different ways, including remittances, buying livestock, gold, or property. Among the diverse options offered, saving cash in a bank account was not considered a form of active saving. In subsequent surveys, "building a balance in a current account" was dropped as it is not regarded as active saving; new options were added to consider different investment forms, including crypto-assets)

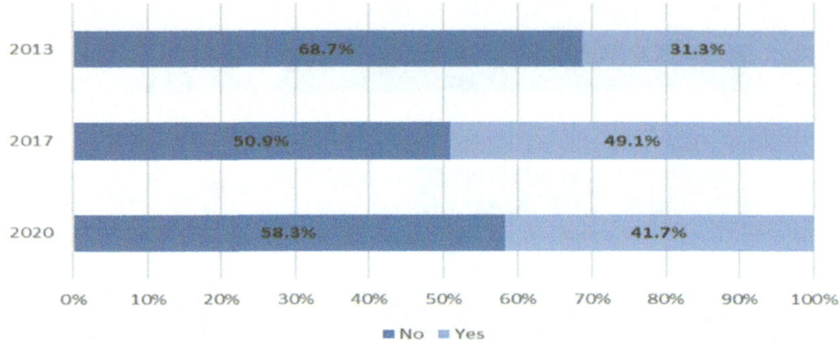

Fig. 7.17 I didn't have negative savings during the last 12 months, and if I did, I didn't borrow to make ends meet (*Note* Figure reports answers given by respondents who had positive savings or did not borrow to make ends meet in case of negative savings)

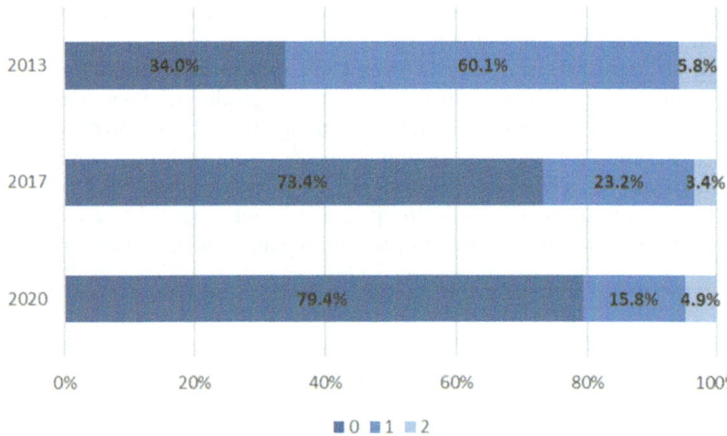

Fig. 7.18 Financial product choice (*Note* Figure reports answers by respondents to the question about financial product choice. In particular, they are asked whether or not they have acquired information to make an informed buy [1 point for people who had tried to shop around or gather any information; 2 points for those who had shopped around and gathered independent information; and 0 all other cases])

Annex 3

CART analysis on the components of financial literacy index

1. Financial Knowledge Index

See Fig. 7.19 and Table 7.9

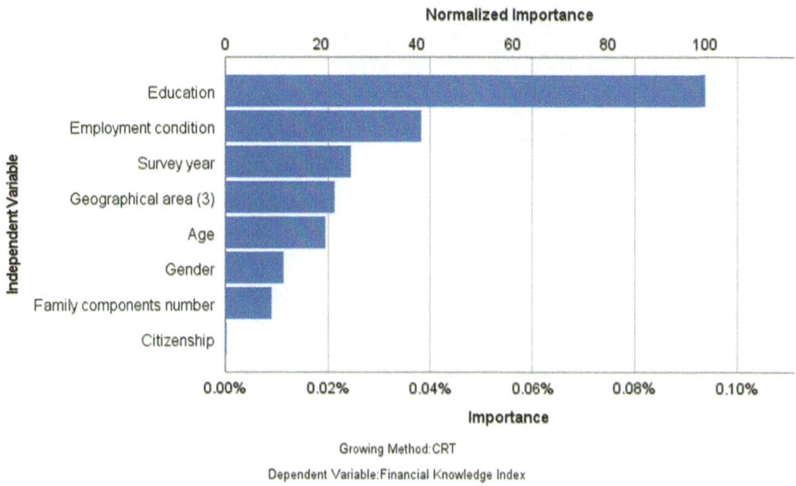

Fig. 7.19 Importance of the independent variables for the construction of the FKI tree

Table 7.9 Description of the 18 final groups by the regression tree on FKI

Cluster	*n*	*Std. Dev.*	*Mean*	*Description*
Cluster 1	1666	1.340	2.316	Education: ≤ Secondary school; North, Centre Italy
Cluster 2	1529	1.466	2.586	Female; Education: > Secondary school
Cluster 3	266	1.260	1.504	Education: Primary school, South Italy
Cluster 4	639	1.318	2.049	Education: Secondary school, South Italy
Cluster 5	339	1.332	3.292	Year: 2013; Male; Education: > Secondary school
Cluster 6	1210	1.507	2.759	Year: 2017/2020; Male; Education: > Secondary school

2. Financial Attitude Index
See Fig. 7.20 and Table 7.10

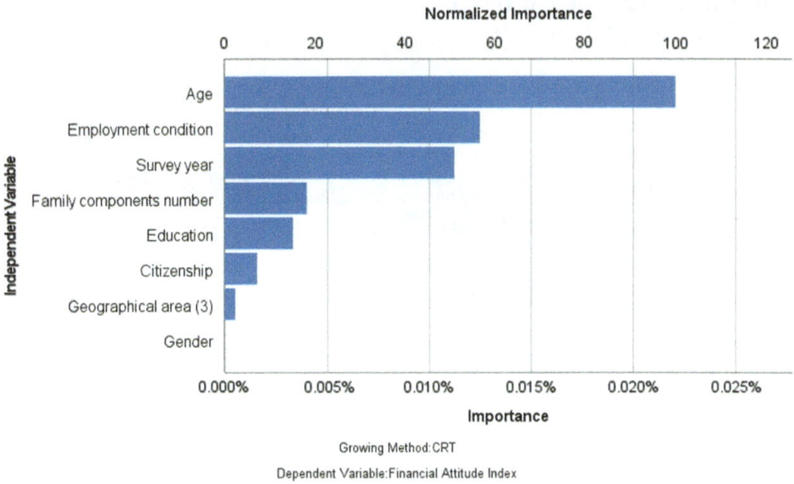

Fig. 7.20 Importance of the independent variables for the construction of the FAI tree

Table 7.10 Description of the 11 final groups by the regression tree on FAI

Cluster	*n*	*Std. Dev.*	*Mean*	*Description*
Cluster 1	567	0.845	3.416	Year: 2013; Education: > Secondary school
Cluster 2	439	0.938	3.308	Year: 2013; Male; Education: ≤ Secondary school; *n*. family component: ≤ 3 persons
Cluster 3	150	0.980	2.967	Year: 2013; Male; Education: ≤ Secondary school; *n*. family component: > 3 persons
Cluster 4	381	0.927	2.675	Year: 2017/2020; Age: ≤ 24
Cluster 5	1261	0.813	2.973	Year: 2017/2020; Age: [25, 44]
Cluster 6	1459	0.812	3.085	Year: 2017/2020; Age: [45, 64]
Cluster 7	1060	0.837	3.266	Year: 2017/2020; Age: > 64

3. Financial Behavior Index
See Fig. 7.21, Tables 7.11 and 7.12

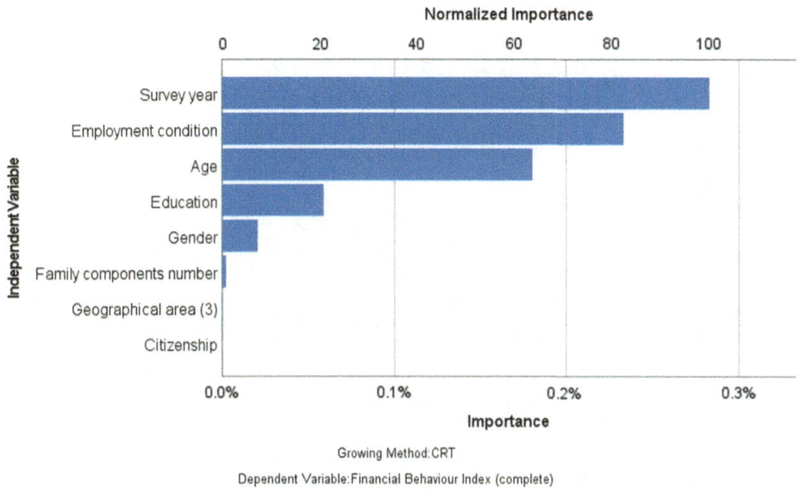

Fig. 7.21 Importance of the independent variables for the construction of the FBI tree

Table 7.11 Description of the 11 final groups by the regression tree on FBI

Cluster	*n*	*Std. Dev.*	*Mean*	*Description*
Cluster 1	495	1.393	6.240	Year: 2013; Employed, Self employed
Cluster 2	696	1.401	5.435	Year: 2013; Retired, Student, Housewife, Looking for a job, Other work conditions
Cluster 3	340	1.797	2.991	Year: 2017/2020; Student/Looking for a job; Age: ≤ 24
Cluster 4	339	1.802	3.976	Year: 2017/2020; Student/Looking for a job; Age: > 24
Cluster 5	1729	1.666	4.533	Year: 2017/2020; Employed, Self Employed, Housewife, Retired, Other work conditions; Education: ≤ Secondary school
Cluster 6	2050	1.819	5.020	Year: 2017/2020; Employed, Self Employed, Housewife, Retired, Other work conditions; Education: > Secondary school

Table 7.12 Characteristics of clusters

Cluster		Gender		Total	Age				Total	New employment condition			Total
		Female	*Male*		*18–24*	*25–44*	*45–64*	*≥ 65*		*Worker*	*Not looking for a work, not worker*	*Looking for a work*	
Cluster 1	*n*	359	264	623	10	162	222	228	622	189	387	47	623
	%	57.6	42.4	100.0	1.6	26.0	35.7	36.7	100.0	30.3	62.1	7.5	100.0
Cluster 2	*n*	290	335	625	102	262	184	78	626	338	232	56	626
	%	46.4	53.6	100.0	16.3	41.9	29.4	12.5	100.0	54.0	37.1	8.9	100.0
Cluster 3	*n*	346	359	705	341	258	100	6	705	0	337	367	704
	%	49.1	50.9	100.0	48.4	36.6	14.2	0.9	100.0	0.0	47.9	52.1	100.0
Cluster 4	*n*	1041	964	2005	37	806	868	295	2006	1500	505	0	2005
	%	51.9	48.1	100.0	1.8	40.2	43.3	14.7	100.0	74.8	25.2	0.0	100.0
Cluster 5	*n*	600	532	1132	12	150	344	626	1132	382	750	0	1132
	%	53.0	47.0	100.0	1.1	13.3	30.4	55.3	100.0	33.7	66.3	0.0	100.0
Cluster 6	*n*	305	264	569	10	113	203	244	570	192	377	0	569
	%	53.6	46.4	100.0	1.8	19.8	35.6	42.8	100.0	33.7	66.3	0.0	100.0
Total	*n*	2941	2718	5659	512	1751	1921	1477	5661	2601	2588	470	5659
	%	52.0	48.0	100.0	9.0	30.9	33.9	26.1	100.0	46.0	45.7	8.3	100.0

Cluster		*Geographical area*			*Total*	*Family component*				*Total*
		North	*Centre*	*South*		*1.00*	*2.00*	*3.00*	*≥ 4.00*	
Cluster 1	*n*	301	107	214	622	154	172	133	162	621
	%	48.4	17.2	34.4	100.0	24.8	27.7	21.4	26.1	100.0
Cluster 2	*n*	279	142	204	625	84	129	185	227	625
	%	44.6	22.7	32.6	100.0	13.4	20.6	29.6	36.3	100.0
Cluster 3	*n*	267	114	324	705	40	86	231	348	705
	%	37.9	16.2	46.0	100.0	5.7	12.2	32.8	49.4	100.0
Cluster 4	*n*	993	405	608	2006	287	561	556	601	2005
	%	49.5	20.2	30.3	100.0	14.3	28.0	27.7	30.0	100.0
Cluster 5	*n*	768	365	0	1133	279	471	198	184	1132
	%	67.8	32.2	0.0	100.0	24.6	41.6	17.5	16.3	100.0
Cluster 6	*n*	0	0	569	569	77	207	105	180	569
	%	0.0	0.0	100.0	100.0	13.5	36.4	18.5	31.6	100.0
Total	*n*	2608	1133	1919	5660	921	1626	1408	1702	5657
	%	46.1	20.0	33.9	100.0	16.3	28.7	24.9	30.1	100.0

References

Angrisani, M., Burke, J., Lusardi, A., & Mottola, G. R. (2020). *The stability and predictive power of financial literacy: Evidence from longitudinal data* (NBER Working Paper No. w28125).

Atkinson, A., & Messy, F. (2012). *Measuring financial literacy: Results of the OECD/International Network on Financial Education (INFE) pilot study* (OECD Working Papers on Finance, Insurance and Private Pensions, No. 15). OECD Publishing.

Borodich, S., Deplazes, S., Kardash, N., & Kovzik, A. (2010). Comparative analysis of the levels of financial literacy among students in the US, Belarus, and Japan. *Journal of Economics and Economic Education Research, 11*(3), 71–86.

Breiman, L., Friedman, J., Olshen, R., & Stone, C. (1984). *Classification and regression trees*. Chapman and Hall, Wadsworth.

Cucinelli, D., Trivellato, P., & Zenga, M. (2019). Financial literacy: The role of the local context. *Journal of Consumer Affairs, 53*(4), 1874–1919.

De Beckker, K., De Witte, K., & Van Campenhout, G. (2020). The role of national culture in financial literacy: Cross-country evidence. *Journal of Consumer Affairs, 54*(3), 912–930.

di Salvatore, A., Franceschi, F., Neri, A., & Zanichelli, F. (2018). *Measuring the financial literacy of the adult population: the experience of Banca d'Italia*. QEF N. 435.

FINRA Foundation. (2019). *National financial capability study*. https://www.finrafoundation.org/sites/finrafoundation/files/NFCS-Report-Fifth-Edition-July-2022.pdf

Lusardi, A. (2019). Financial literacy and the need for financial education: Evidence and implications. *Swiss Journal of Economics and Statistics, 155*(1), 1–8. https://doi.org/10.1186/s41937-019-0027-5

Lusardi, A., & Mitchell, O. S. (2014). The economic importance of financial literacy: Theory and evidence. *Journal of Economic Litterature, 52*(1), 5–44.

Lusardi, A., Mitchell, O. S., & Curto, V. (2010). Financial literacy among the young. *Journal of Consumer Affairs, 44*(2), 358–380.

Mazzonna, F., & Peracchi, F. (2020). *Are older people aware of their cognitive decline?* (Misperception and Financial Decision Making, IZA DP No. 13725).

Nicolini, G., Cude, B. J., & Chatterjee, S. (2013). Financial literacy: A comparative study across four countries. *International Journal of Consumer Studies, 37*(6), 689–705.

OECD. (2015). *National strategies for financial education*. OECD/INFE Policy Handbook.

OECD. (2016). *OECD/INFE international survey of adult financial literacy competencies*. OECD. www.oecd.org/finance/OECD-INFE-International-Survey-of-Adult-Financial-Literacy-Competencies.pdf

OECD. (2020). *OECD/INFE 2020 international survey of adult financial literacy*. www.oecd.org/financial/education/launchoftheoecdinfeglobalfinancialliteracysurveyreport.htm

OECD/INFE. (2018). Toolkit for measuring financial literacy and financial inclusion.

Schmeiser, M. D., & Seligman, J. S. (2013). Using the right yardstick: Assessing financial literacy measures by way of financial well-being. *Journal of Consumer Affairs, 47*(2), 243–262.

CHAPTER 8

An Interdisciplinary Approach to Economic Texts: The "Considerazioni Finali" by the Governor of the Bank of Italy as a Case Study

Paola Vezzani, Cristina Guardiano, and Valentina Ligabue

8.1 Introduction

Every year, at the end of May, the Bank of Italy publishes a *Relazione Annuale* (Annual Report, henceforth RA). A synthesis of the content of this document (called *Considerazioni Finali*, henceforth CF) is presented by the Governor of the Bank, in a public meeting, a formal event that has been held every year in May from 1947 until now. The CF

P. Vezzani (✉) · C. Guardiano · V. Ligabue
Department of Communication and Economics, University of Modena and Reggio Emilia, Viale Allegri, 9, Reggio Emilia, Italy
e-mail: paola.vezzani@unimore.it

C. Guardiano
e-mail: cristina.guardiano@unimore.it

P. Wachtel et al. (eds.), *Creating Value and Improving Financial Performance*, Palgrave Macmillan Studies in Banking and Financial Institutions, https://doi.org/10.1007/978-3-031-24876-4_8

contain a comprehensive analysis of the key-events concerning national and international economy.

In this chapter, we propose an analysis of the 75 CF pronounced from 1947 to 2021. The goal of our research is to investigate the relation between the content of these documents and their linguistic nature, to check whether and how the latter instantiates the technical features of typical economic texts while, at the same time, reflecting economic, social, and cultural changes happened to the country over the past century. Our analysis combines a detailed survey of the content of the 75 CF, carried out using the tools of economic analysis, with an observation of the linguistic configuration of these texts. The CF display at least three properties which are relevant for these purposes: (1) they define a detailed picture of the major historical events that have shaped the recent history of Italy, along with their consequences on the social and economic sphere; (2) they are a reliable indicator of the changes occurred to our country not only in terms of social and economic life but also in terms of its linguistic identity; (3) while they are sufficiently "flexible" documents to absorb and reflect the major changes which have shaped the recent history of Italy, at the same time they also instantiate a rigid linguistic and stylistic format that depends on their official status as institutional documents.

Our approach combines methods and goals of two usually divergent disciplines (economics and linguistics), to check whether their interaction provides insights which would not be reached through a monodisciplinary lens. We focus on the relationship between the technical content of the CF and specific lexical choices. We single out the most significant economic, financial, and social events at the national and international level, to trace back the economic history of Italy from the post-war period to current times, and we explore the major trends in the usage of the technical items associated with such a content.

The chapter is structured as follows. Section 8.1 introduces the topic and explains the structure of the study. Section 8.2 defines some general background concerning the nature and role of the CF within the Italian economic and financial scenario, presents some statistics regarding these documents, sketches a brief sociolinguistic background, and summarizes some relevant literature. Section 8.3 summarizes the contents of the CF, singles out some keywords, and provides some statistical information. Section 8.4 discusses some linguistic features. Section 8.5 summarizes the preliminary results.

8.2 Setting the Stage

The CF were introduced in 1946 and are a reference point for experts in economy and finance. Table 8.1 summarizes some information about each Governor, their mandate, and their CF. Table 8.2 summarizes their textual features. Figure 8.1 visualizes the distribution of their length. The full text, also available in English, of all the CF pronounced so far can be found at https://www.bancaditalia.it/pubblicazioni/relazione-annuale/index.html?com.dotmarketing.htmlpage.language=1

8.3 The Position of the CF in the Sociolinguistic Architecture of Italian

According to a "classical" sociolinguistic perspective (Berruto, 1995; Chambers, 2003; Labov, 1972, 2001), synchronic linguistic diversity follows from four major sets of variable factors, traditionally labeled "dimensions of variation", and briefly summarized in Table 8.3.

Although the Governors come from different areas of Italy (Table 8.1), the *diatopic* dimension is not salient in their texts: the CF are formal, specialized texts and, as such, they display the linguistic features typical of written and technical varieties of Italian, which are strongly standardized and, as such, do not display any local/regional peculiarity. Concerning the *diaphasic* dimension, the CF are highly specialized texts: thus, they adopt the scientific and technical variety typical of economic and financial documents. The event in which they are presented is institutionalized, with no flexibility in terms of individual choices. Concerning the *diastratic* dimension, the authors of the CF (the Governor and his staff) are specialists, with a high level of education and professional competence, as well as a high degree of experience in the fields of economics, banking, and finance. As such, they master the formal, specialized language varieties associated with their area of expertise. The *diamesic* dimension is one of the most interesting for the purposes of the present work. The CF are conceived, and actualized from the very beginning, as written texts. They are presented by the Governor in a formal speech; hence, they are pronounced orally, but with no difference with respect to the written text. The audience, while attending the reading, is provided with a written copy of the text: hence, these documents are closer to typically written texts than to spoken ones.

Table 8.1 The Governors of the Bank of Italy and their CF

Governor	*Mandate*	*Length of the mandate (days, years)*	*Age at the beginning of the mandate*	*Number of CF*	*Years*
Bonaldo STRINGHER (Udine, December 1854–December 1930)	July 3, 1928–December 24, 1930	907 2.5	73	Not present	–
Vincenzo AZZOLINI (Napoli, December 1881–August 1967)	January 10, 1931–June 4, 1944	4894 13.4	49	Not present	–
Luigi EINAUDI (Carrù, Cuneo, March 1874–October 1961)	January 5, 1945–May 11, 1948	1222 3.3	70	1	1946
Donato MENICHELLA (Biccari, Foggia, January 1896–July 1984)	August 7, 1948–August 17, 1960	4393 12.0	52	13	1947–1959
Guido CARLI (Brescia, March 1914–April 1993}	August 18, 1960–August 18, 1975	5478 15.0	56	15	1960–1974

Governor	*Mandate*	*Length of the mandate (days, years)*	*Age at the beginning of the mandate*	*Number of CF*	*Years*
Paolo BAFFI [Broni, Pavia, August 1911–August 1989)	August 19, 1975–October 7, 1979	1510 4.1	64	4	1975–1978
Carlo Azeglio CIAMPI (Livorno, December 1920–September 2016)	October 8, 1979–April 29, 1993	4952 13.6	58	13	1979–1991
Antonio FAZIO (Alvito, Frosinone, October 1936)	May 4, 1993–December 20, 2005	4613 12.6	56	13	1992–2004
Mario DRAGHI (Roma, September 1947)	December 29, 2005–October 31, 2011	2132 5.8	58	6	2005–2010
Ignazio VISCO (Napoli, November 1949)	November 1, 2011–*in progress*	3713 10.2	62	10	2011–2020

Table 8.2 Descriptive statistics (mean values)

Governor	*# of CF*	*# of Words*	*# of Pages*	*# of Sections*	*# of Subsections*	*Annex (tables and figures)*
EINAUDI	1	9.946	36.0	–	–	Yes
MENICHELLA	13	11.484	36.7	3.2	–	–
CARLI	15	14.512	41.3	4.5	–	–
BAFFI	4	16.084	41.8	7.5	–	–
CIAMPI	13	12.844	32.6	4.9	4.5	–
FAZIO	13	11.042	31.9	3.7	10.2	–
DRAGHI	6	6.693	18.0	3.0	4.7	–
VISCO	10	8.805	23.9	4.6	0.8	Yes (from 2016)
TOTAL	75	11.733	33.1	4.2	3.0	

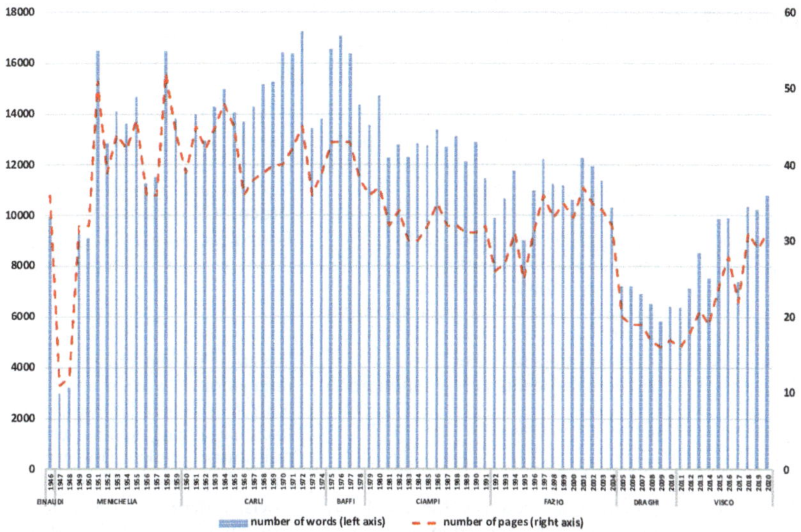

Fig. 8.1 The length of the CF

From the point of view of their lexical features, one major trait of specialized varieties is the bi-univocal correspondence between form and meaning (Saussure, 1916). Synonymy and polysemy are excluded from their lexicon: each word has a specialized meaning and corresponds to a specific notion/definition associated with a given technical content (Cortelazzo, 1994). Typical examples of technical terms, in the specialized variety of economy and finance in Italian language (see Dardano, 1998: 65–69; Rainer, 2006: 2148–2149; and, for an overview, Proietti, 2010), are "inflazione" (inflation) or "derivato" (derivative). Several such items are known to a non-specialized audience, which yet has often just a vague idea of their meaning, ignoring the technical details of their definition.

In specialized varieties, technical terms are often borrowed from other languages: terms like "private equity" or "asset management" are used in Italian as borrowings from English (Rando, 1990). A further mechanism to generate technical terms is the resemantization of words originally belonging to the common lexicon (e.g., "vigilanza/supervisory", "azioni/stock", etc.; Rainer, 2006) or to the specialized lexicon of other disciplines (Cortelazzo, 1994). Specialized lexicon is often rich in

Table 8.3 Dimensions of variation

Dimension	*Definition*	*Linguistic manifestations*
Diatopic	Variation depending on the geographic distribution of the speaker communities and the linguistic history of individual speakers	Different local varieties of one and the same language (e.g. regional varieties of Italian), Multilingual vs. plurilingual structure (Edwards, 1994)
Diaphasic	Variation depending on different communication situations or settings, or the relationship between the speaker involved in a given linguistic event	Formal vs. informal registers, different types of literary styles, etc
Diamesic	Variation depending on the medium/modality of communication	Varieties: spoken (face to face interactions}, written, digital written (produced through tools like chat, email, texting, etc.), transmitted spoken (for instance through a Skype call), etc
Diastratic	Variation according to the social class or group a speaker belongs to, to his/her cultural background, etc. This dimension defines the structure of the speaker's competence through the investigation of various demographic and cultural variables (Hymes, 1966), and how it is 'externalized' in actual language use	'social varieties' of language: from highly educated speakers to working-class varieties

acronyms (e.g., PIL, Prodotto Interno Lordo, Gross Domestic Product; CET1, Common Equity Tier 1; BCE, Banca Centrale Europea, European Central Bank, etc.).[1] Table 8.4 lists some examples of technical terms (first column), along with their definition (second column).

The use of a specialized lexicon has two advantages: it excludes ambiguity (to guarantee objectivity in the definition of theories and data) and favors synthetic style. To give a practical example, we provide in Table 8.5 some excerpts from the CF containing technical terms (first column) with a potential non-technical "translation", where no technical terms are

[1] A list of acronyms used in the lexicon of economy and finance in Italian can be found here: https://www.bancaditalia.it/pubblicazioni/relazione-annuale/2020/app_2020_totale.pdf.

Table 8.4 Technical terms

Technical terms	*Technical definition*
Eurosistema (Eurosystem)	The Eurosystem comprises the European Central Bank(ECB) and the National Central Banks (NCBs) of those countries that have adopted the euro. The Eurosystem and the European System of Central Banks (ESCB) will co-exist as long as there are EU Member States outside the euro area *Source* https://www.ecb.europa.eu/services/glossary/html/glossg.it.htmI
Indice di stabilità della provvista a medio e lungo termine (Net stable funding ratio–NSFR)	The amount of available stable funding relative to the amount of required stable funding. Available stable funding is the portion of capitaI and liabilities that is expected to be stable over a one-year time horizon. The amount of funding required of a specific institution is a function of the liquidity characteristics and residual maturities of the various assets held by that institution as well as those of its off-balance sheet (OBS) exposures. The NSFR should be equal to or higher than 100% *Source* https://www.ecb.europa.eu/services/qlossary/html/glossg.it.html
PIL–Prodotto Interno Lordo (GDP–Gross Domestic Product)	A measure of economic activity, namely the value of an economy's total output of goods and services, less intermediate consumption, plus net taxes on products and imports, in a specified period. GDP can be broken down by output, expenditure or income components. The main expenditure aggregates that make up GDP are household final consumption, government final consumption, gross fixed capital formation, changes in inventories, and imports and exports of goods and services (including intra-euro area trade) *Source* https://www.ecb.europa.eu/services/glossary/html/glossg.it.html

employed (second column). As it can be seen even at a first sight, the non-technical version is much longer and much less precise than the technical one.

The texts analyzed in this chapter have been produced across a time span of more than seventy years. Over these years, the Italian language and its speakers have undergone a huge number of significant transformations, such that there is no linguistic manifestation (from literary texts to everyday speech) that has not changed: we do not speak the same language as 75 years ago, and many speakers (especially the youngest generations) are not able to process the varieties of Italian produced 75 years ago. Against this background, it is quite surprising that, except for obvious lexical differences (described in Sect. 8.3), no salient change is visible in the textual structure, syntax, and style of the CF (Ligabue, 2021). This is due to their rigid diaphasic nature, which has prevented them from massive transformation.

The lexicon is patently affected by conscious individual choices. Yet, languages are not made of words. Every time speakers produce a message, they build up a *structure*, where each word plays a specific function and

Table 8.5 Technical vs. non-technical varieties

Excerpt	*Non-technical "translation"*
1. Ciampi, CF1991 "**L' aumento dei prezzi** elevato ovunque, ha assunto in alcuni paesi caratteri di *iperinflazione*" "**The rise in prices,** which is rapid throughout the region, has turned to *hyperinflation* in some counties"	Prices, in all sectors and all countries, are substantially rising. In some countries they are tremendously rising and are now out of control
2. Fazio, CF2000 "Le entrate risentiranno del **rallentamento congiunturale** e del **ristagno del** *corsi azionari* **registrato nel 2000**" "Revenue **will** be affected by the **cyclical slowdown in growth** and **last year's stagnation in** *equity prices*"	Italy will get a smaller amount of public money, because the economy of our State is currently slowing down. Additionally, in 2000, the stock prices have not increased: this will further reduce the amount of money which our State will be able to use for public purposes
3. Draghi, CF2008 "II **deterioramento dell'economia** tende *a* frenare i ***prestiti bancari***" "The **deterioration of the economy** tends to curb *bank lending*"	When the economy of a country gets worse, firms and families stop asking bank loans; banks, in turn, are less prone to allow loans to both firms and families

Note *Italics* signal technical terms/locutions. **Boldface** characters indicate ***nominalizations*** contributes to the meaning of the whole utterance. Structures define a deeper, more abstract, and less visible level of language. As such, they are less prone to be consciously manipulated by the speakers (e.g., for purposes of communication effectiveness), and are more stable across space and time.

One of the structural features most frequently associated to specialized varieties is "syntactic density", i.e., the presence of compact hierarchical structures with a high density of technical information. This is obtained through several strategies. The first is hypotaxis (subordination), as opposed to parataxis (coordination): technical texts prefer hypotaxis, but according to Astuti et al. (2020: 54) in the CF, parataxis is more frequent than hypotaxis. The second strategy is nominalization, i.e., using a noun derived from a verb in the place of a full sentence. For example, the noun phrase "John's description of the accident" contains the noun *description* and two complements expressing (i) the person who made the description (i.e., the agent: *John*), and (ii) the event that was described (*the accident*). This equals the sentence "John describes the accident", where the subject (*John*) expresses the agent, and the object (*the accident*) expresses the event described. Examples of these structures in the CF are listed in Table 8.4[2]: in (1), "aumento dei prezzi" (rise in prices) corresponds to "i prezzi sono aumentati" (prices have risen); in (2), "rallentamento congiunturale e ristagno dei corsi azionari" (cyclical slowdown and stagnation in equity prices) corresponds to "equity prices have slowed down cyclically and have stagnated"; in (3), "deterioramento dell'economia" (deterioration of the economy) corresponds to "economy has deteriorated". The third strategy is "deagentivization". In the sentence "John describes the accident", the subject is the agent and must be visible. By contrast, in the noun phrase "the description of the accident", the agent is (ad can be) omitted. The same effect of "agent avoidance" is obtained through passivization (Gotti, 1991), the use of passive verbs instead of active ones. In the sentence above, the verb "describes" is in the active voice: this means that the subject (*John*) expresses the agent, while the object (*the accident*) expresses the event described. In its passive counterpart ("the accident was described by John"), the subject expresses the event described, there is no object, and the agent is realized as a prepositional complement, which can be omitted. Finally, a "depersonalization"

[2] For more examples, see Ligabue (2021).

strategy is avoiding first person, which obeys the need of focusing on facts, data, and hypotheses rather than individuals. In the CF, first person is used deictically (it refers to the people actually present in the context) both in the singular (it refers to the Governor and usually signals he is taking direct responsibility) and in the plural (it either refers to the Bank of Italy or inclusively refers to the speaker and the audience).

(1) "*I would like to add a short mention concerning organizational human aspects which have allowed and allow an efficiency that **we** believe **we** must be proud of*" [Carli, CF 1970: 2 in the Italian version. This page was not available in the official English translation.].
(2) ***We**, as the intermediaries between the sources of savings and production, are faced, and will continue to be faced, by difficult choices in the guidance of monetary flows*" [Carli, CF 1970: 44].
(3) "*Today, in a more difficult international environment but one that is also rich in opportunities, **we** must find renewed dynamism, with all sections of society contributing. The post-war economic miracle can be repeated. **We** can and must achieve it*" [Fazio, CF 2000: 32].

8.4 Some Relevant Literature

Works about the CF are of two types. Most of them focus on the vision and thoughts of the Governors. For example, Gai (1995) proposes an analysis of the role of the CF as a moral suasion tool; Valente (1990) traces the first 40 years of CF from Einaudi to Ciampi (1947–1986). Also, several works celebrate the action of individual Governors: Barucci (2008), Patuelli (2016b) Savona (2008) and Signorini (2021) for Einaudi; Bianchi et al. (1998) and Patuelli (2016a) for Menichella; Fazio (1993, 2003) and Gigliobianco (2019) for Carli; Sarcinelli (1999), 2013, Gigliobianco and Massaro (2010), Savona (2013) and Barbiellini Amidei (2020) for Baffi; Peluffo (2007), Ciocca (2018a, b) for Ciampi; Alberici (2006) for Draghi.

Two recent works (Astuti et al., 2020; Bruno, 2016) provide a completely different angle: they use the CF to exploit quantitative analyses based on lexical data extraction and statistical manipulation. In what follows, we summarize their major points in relation to our work.

Bruno (2016) explores the connection between the lexical choices and the content of 20 CF (1996–2015) through sentiment analysis and text mining, to explore the potential of "some main methodologies employed in text mining and for the extraction of sentiment and emotions from textual sources" (Bruno, 2016: 1700). He selects a set of highly frequent words (*banca, mercato, capitale, credito, lavoro, rischio*) which, as will be shown in 8.5.6 also appear in the 50-keyword list singled out from our corpus. Then, he discusses word clouds, showing that "the focus of the words has shifted from the banks in 2005 to the firms and the State in 2010 to arrive, in 2015, to financial system, crises and banks" (Bruno, 2016: 1703). Using sentiment analysis, he extracts general trends concerning the attitudes of the Governors: he shows that "these documents tend to stay pretty much neutral over all the extension of the speech" (Bruno, 2016: 1708), a conclusion that, along with the high average level of the formality score, is consistent with the technical/specialized nature of these documents. He concludes that "the objectivity of these figures provides sound grounds for increasing the accountability and transparency of central bank communications".

Astuti et al. (2020) investigate a corpus of 73 CF (1946–2018). With the aid of word clouds, they explore the distribution and frequency of selected words and locutions and define the major trends emerging from the lexical choices of the Governors. They explore lexical changes, measure the legibility index of the CF using the average length of the sentences, and list the types of subordinate clauses found in the texts.

8.5 Towards an Analysis

8.5.1 Methodology

In this Section, we observe the lexical choices (i.e., words and locutions, which we call *keywords*) that are associated with the contents of the CF, which in turn reflect the national and international economic scenario at the time in which each text was produced. As already remarked, these documents instantiate a special type of language variety that is called "specialized language". The linguistic structure of specialized languages differs from non-specialized varieties in several respects. For example, in non-specialized varieties, lexical choices are heavily determined by the content/information that speakers want to share and by other types of restrictions depending on language usage and speakers' "habits". By contrast, in specialized varieties, word selection is not a free choice of individual speakers, because it is constrained by the specialized nature of the content that must be conveyed. Each word is selected uniquely on the basis of its technical meaning, rather than on the basis of any other type of stylistic or individual considerations.

To better explore this aspect, in the following Section, for each decade and/or Governor, we first present a summary of the major events that occurred to the national and international economy during his mandate (box), then we list some excerpts of the CF which are representative of those events, and finally, we discuss the 50 keywords (mostly technical terms) that are associated with each specific content-block. The keywords were selected "manually" based on the analysis of the content of each individual CF. This type of analyses are usually conducted through the aid of automatic tools for lexical data extraction (see for instance Bruno, 2016; Astuti et al., 2020), which have the purpose of exploring the structure of the content on the basis of the distribution of lexical choice. We adopted an opposite perspective, and started from a qualitative analysis of their content, that we subsequently used as a guide to select the list of keywords and to explore their distribution. The latter is described through the aid of descriptive statistics in Sect. 8.5.3.

The 50 keywords summarize, mainly through technical terms, the main contents of the CF, and therefore they are indirect examples of the orientations of the Italian economy in relation to the national and international events and contexts from which each text originates. In identifying the keywords, we selected a representative set of the major general issues addressed in the CF, excluding single specific contingencies.

8.5.2 Pills of Italian Economic History

After the War and Luigi Einaudi (January 1945–May 1948)

At the end of World War II Italy was in a dramatic state, having to deal with a sunken economy characterized by very high inflation, by the devaluation of its currency (lira) and by prices which exceeded pre-war ones by 20-fold.

The Bank of Italy, together with other institutions, faced difficult times. In January 1945, Luigi Einaudi was appointed Governor. He became a key figure during these first years of reconstruction, because he guided the country towards a progressive way out of the economic crisis. In a country worn out by the war, Einaudi succeeds in making a close link between economic policy choices and long-term strategies. His persuasive thoughts were presented in the first ever CF (concerning the year 1946 and presented in 1947). In 1948 he was then elected President of the Republic.

One of the main issues in CF 1946 by Einaudi was the trend of ***prices (prezzi):*** the Italian economy was characterized by very high ***inflation (inflazione),*** whose negative effects reflected on ***savings (risparmio/i).*** These were a key factor for the development of the country: "*...**saving is a function of confidence in the monetary unit***" [CF 1946, 20]. Thus, the main goal of monetary policy was price stability. The term ***money (moneta)*** is frequently used as well.

The "Economic Miracle" and Donato Menichella (August 1948–August 1960)

The post-war period was the so-called 'economic miracle': between 1951 and 1963, Italian underwent the deepest transformation in its contemporary history.

The key factor which triggered economic development was the gradual opening-up to the international economy. The growth of exports induced the strengthening of competitiveness and improvement of the industrial system. Investments, industrial productivity, international trade, competitiveness of Italian companies and gross domestic product rapidly increased. This attitude was consolidated in 1957 through Italy's entry into the European Economic Community.

When Menichella left the Bank of Italy in 1960, the lira was strong and stable in terms of purchasing power in comparison to other currencies.

Donato Menichella's mandate lasted 12 years. Like Einaudi, he focused on *prices (prezzi)* and *savings (risparmio/i);* less frequent in his CF are the terms *inflation (inflazione),* since it had been reduced, and *money (moneta)*. Other important issues in his CF are *foreign exchange reserves (riserve),* which have strategic importance to favor savings, *and industrial production (produzione industriale):* Italy achieved positive results, and the maintenance of high production levels was the necessary goal to foster *monetary stability (stabilità monetaria)*. For Menichella, the goal to be pursued is *development (sviluppo),* which depends on *investments (investimenti),* which in turn depends on savings: the use of the keyword "savings" reaches high frequencies, only lower than Ciampi's 1988 CF.

Raw materials (materie prime) were also a topic of interest, mainly due to supply difficulties in many countries.

Menichella also dealt extensively with the functions and activities carried out by the *credit system (sistema creditizio):* the keyword *credit (credito)* appears 28 times per CF. In 1957, the keyword *innovations (innovazioni)* appeared for the first time, linked to the reduced innovative development of some industry sectors. The term *expansion (espansione)* is also frequent in his CF.

Stormy Years and Guido Carli (August 1960–August 1975)

Unlike the years of the 'economic miracle', which had been characterized by an equilibrium in the balance of payments and monetary stability, the 1960s were instead marked by very critical conflictual situations that reached a climax with the union struggles of 1962–1963s and 1969–1970s.

The rise in the employment rate, especially in northern Italy, caused an increase in wages, followed by a generalized rise in prices. The different contradictions that had characterized the growth of the 1950s favored the development of tensions.

In the 1970s, Italy also suffered the consequences of US-based political and economic decisions: to cope with the growing debt of the United States, President Nixon started the flexible exchange rate system by interrupting the Bretton Woods agreements. This monetary policy had brought significant advantages to Italy, as it had allowed a better control of inflation and a greater stability of the financial system, and moreover, thanks to the surplus in the balance of payments, Italy was a country with a strong currency. Among the relevant consequences of the end of the Bretton Woods agreements, there was a high devaluation of the lira against the dollar, which led to an increase in the price of raw materials and imported goods.

Carli paid higher attention to the financial *market (mercato)* and to *monetary policy (politica monetaria)*. In addition, Carli frequently mentions *innovations (innovazioni)*, which are useful in creating new tools for businesses and new *technologies (tecnologie)*, and first appeared in 1964. During the 1960s, the Governor focused on the *banking system (sistema bancario)*, *banks* (at the time called "*aziende di credito*") and *credit (credito)*, with particular attention to *interest rates (tassi di interesse)*. As compared to Menichella, attention to *production (produzione)*, *raw materials (materie prime)*, *inflation (inflazione)*, and *monetary stability (stabilità monetaria)* is reduced. The keyword *liquidity (liquidità)* is one of the most frequent in his texts; the *public sector (settore pubblico)* becomes a constant but also *competition (concorrenza)* is more frequent as compared to the past. Carli is the first Governor who emphasized the importance of *expectations (aspettative)* for monetary policy management. Like his predecessor, he highlights the importance of *investments (investimenti)* and *expansion (espansione)*.

The Second Oil Crisis and Paolo Baffi (August 1975–October 1979)

Paolo Baffi became Governor during extremely difficult years, not only for the country's economy, characterized by currency crises, bank crises, political and financial scandals, but also for Italian democracy, marked by Prime Minister Aldo Moro's murder.

The years 1975–1979 were characterized by the negative consequences engendered by the increase in oil prices and by the worsening of domestic production prices caused by imported raw material.

The second half of the 1970s was characterized by the second oil shock, the lira depreciation and the end of Bretton Woods' agreements. Therefore, Paolo Baffi's attention shifted back to *prices (prezzi)*, *inflation (inflation)*, *raw materials (materie prime)*, and *monetary stability (stabilità monetaria)*. In Baffi's CF, the term *credit (credito)* reaches its highest frequency (57 times per CF), due to the attention paid both to *banks (aziende di credito)* and to proper bank management. References to *innovations (innovazioni)*—for fostering competitiveness on international markets and the oil and currency crisis—increased.

Both Carli and Baffi denounced risks resulting from an excessive presence of the *public sector (settore pubblico)* in the Italian economy: as a consequence, *market (mercato)* is a highly frequent word in Baffi's CF. Baffi's discontinuity in supervisory is more pronounced than Carli's, and he provides more information about central bank activity. Baffi is the first who writes, for the year 1976, a section called "The supervision of credit operations". The word *supervisory (vigilanza),* which Menichella and Carli had rarely used, appears twelve times in his 1975 CF and hits a maximum of nineteen times in the 1978 one.

The 1980s, the European Market Union and Carlo Azeglio Ciampi (October 1979–April 1993)

The year 1980 began in a scenario of decreasing development and increasing inflation. From that moment on, in Italy and abroad, the reduction of inflation became the main goal to be pursued. The greatest impulse, which determined the increase in the cost of money within international markets, came from the US, where the Chairman of the FED applied a restrictive monetary policy to fight stagflation and to support an increase in interest rates.

The Italian participation in the European Monetary System (EMS) had some important consequences, such as the transition from Fordist production systems to network systems, as well as increased merger and acquisition operations, resulting in an improvement in profits and a reduction in inflation, which, however, remained higher than in other countries.

Also, in 1981, the well-known "divorce" between the Bank of Italy and the Treasury took place. As a result, the Bank of Italy quit the practice of subscribing short-term government bonds not being sold at auctions.

During Ciampi's mandate, the second oil shock and the slowdown in development forced him to deal mainly with *inflation (inflazione)* and the need for a stable currency. Great attention was given to *monetary policy (politica monetaria),* a keyword widely used by Ciampi. The frequency of *interest rates (tassi di interesse)* increased, especially regarding the reduction of US interest rates, and also the keyword *growth (crescita),* referring to the growth of economy, productivity, product, and money supply, became more frequent. All Ciampi's CF show a high frequency of *innovations (innovazioni):* he describes technical innovations within financial intermediaries, non-financial firms, and also law contents.

Ciampi used the word *money (moneta)* 22 times in 1980 and 27 in 1991, much higher than other Governors; in 1983, he used *public debt (debito pubblico)* 15 times. With Ciampi, the reform of the Italian banking system took place, followed by the branch's deregulation and the confirming idea that the bank is a firm. In 1991, he used *competition (concorrenza)* 21 times, the highest value in the historical series, even though this would later be taken up by Draghi.

The 1990s and the Currency Crisis: From Carlo Azeglio Ciampi to Antonio Fazio (May 1993–December 2005)

The beginning of the 1990s was a stormy period for Italy. In February 1992, the Maastricht's Treaty was signed, thus establishing the European Community. Tensions deriving from the choice of some countries not to adhere to the European Monetary Union brought a serious currency crisis, which blew up in September of the same year when the pound and the lira were devalued and forced out of the European Monetary System. Yields on government bonds increased significantly, further increasing the cost of public debt cost.

After the 1992 crisis, investments gradually improved and, in 1994, Italy registered a GDP's acceleration thanks to the increase of other countries' demand, favored by the devaluation of the lira. At the beginning of 1999, Italy joined the European Monetary Union.

In 1993 (Bank of Italy's centenary), Ciampi resigned because he was appointed Prime Minister, and Antonio Fazio became Governor.

In the 1990s, during the currency crisis, Fazio focused on *interest rates (tassi di interesse)* and on their variations and consequences. In Fazio's CF, we find a high frequency of the term *development (sviluppo)*, which increased especially years later in the 2000s, in relation to the development of stock markets, stock exchange, industrial sector, and other innovative sectors. In addition, Fazio, together with Visco, is the Governor who repeatedly mentioned new *technologies (tecnologie)*. Moreover, after two decades, Fazio came back and gave more attention to the *banking system (sistema bancario)*, dealing both with its deregulation and with *supervisory (vigilanza)* activity.

The economic situation led to new themes. The word *globalization (globalizzazione)* appears for the first time in 1996. As in the previous decade, the concern for the conditions of *public finance/public sector (finanza pubblica/settore pubblico)* was high. The period from 1995 to 2000 was characterized by a sharp rise in stock prices and by the birth of New Economy; the term *stock exchange (borsa)* reached the highest frequency in the history of the CF.

The euro currency became the main subject in the 1997 CF: Fazio pronounced this word 10 times and moreover during his mandate in his CF, he pronounced 85 times the keyword *euro area (area dell'euro)*.

The 2000s and the Financial Crisis: From Antonio Fazio to Mario Draghi (December 2005–October 2011)

During the first decade of the twenty-first century, the Italian economy's growth rate moved further away from the average of the main Euro countries, and productivity halved with respect to the previous decade.

The financial crisis that broke out in 2008 lasted more time in Italy than anywhere else: GDP collapsed, unemployment increased, investments, consumption, ad export fell down, and the number of manufacturing firms decreased.

One of the main reasons for Italy's inability to adapt to the pace of innovation was the structure of the Italian production system: the high number of small and medium enterprises, with scarce financial resources supporting the increasing costs in research and development, was a concrete obstacle to technological progress. The rapid increase in public debt was one of the causes that favored economic stagnation.

Mario Draghi was appointed at the end of 2005, when Fazio resigned. In continuity with Fazio, Draghi gave increasing importance to the *banking system (sistema bancario)*, which was in great difficulty due to the financial crisis and was also undergoing a large reorganization process. All his CF contain a section specifically devoted to this topic, with increasing references to *supervisory (vigilanza)*, related to the reform of the Basel agreement and other law's revisions. The keywords *risk/risks (rischio/i)* and *crisis (crisi)* reached very high frequencies (20 times each per CF). A fundamental topic was the *growth (crescita)*, a goal to be achieved through higher flexibility in the labor market, competitiveness of services, investment in *innovation (innovazione)*, and *capital market's* development *(mercato)*.

Draghi also devoted great attention to *firms (imprese)*, in relation to Italy's loss of competitiveness and fall in productivity. The keyword *country (paese)* appears more than in the past as evidence of his concern to this topic.

Last Ten years and Ignazio Visco (November 2011—Mandate in Progress)

In just five years Italy faced two serious crises. The first, the 2008 financial crisis mentioned above, resulted in a reduction in GDP of about 7% in our country. This recession was followed by a little recovery, which was, however, temporary. Indeed, in the second half of 2011, the sovereign debt crisis exploded.

In our country, the ten years sovereign rate reached 7% and the spread with the German bond increased from 200 to more than 500 basis points. Italy found itself, again, in a situation of reduced economic activity, which caused a drastic increase in unemployment.

The growing stress on the interbank market led the ECB to intervene, implementing refinancing operations with the aim of reducing pressure on the money market. The positive effects were limited, however, as the serious market conditions prevented the ECB's stimulus from being reflected effectively on the real economy. In 2015, the ECB started its quantitative easing, through which the purchase of assets of commercial banks started, to support economic growth and reduce inflation. This intervention resulted in a significant reduction in sovereign debt yields and, therefore, in a lower perception of the risk at European level.

In 2016, the Italian economy was in expansion: improvements happened in various sectors, from industry and services, from exports to investments. Throughout 2017, the economy kept strengthen, recording product growth of 1.5%, a significant increase in employment and a considerable expansion of production in all sectors. Slowdowns, in terms of growth and development, characterized 2018, when uncertainties and the slowdown in German activity negatively affected Italy, which recorded a growth of only 0.9%. The result was a reduction in investments, consumption, and employment.

The most recent history worldwide, starting from the end of 2019, sees the Covid-19 pandemic as an epochal event with an extremely severe impact on all levels. The spread of the SARS-CoV-2 virus swept through the whole global economy. With many epidemic waves, the economic effects played out differently across sectors and geographical areas, reflecting the severity of the pandemic at local level and the economic policy responses.

Monetary policies prevented the pandemic crisis from morphing into a financial crisis by guaranteeing liquidity to the markets and facilitating credit through various measures, including asset purchase programs. Euro-area GDP recorded the sharpest contraction since the inception of the Monetary Union.

Last year Italy's GDP recorded its largest drop since World War II (−8.9%) and this big impact arose from different causes: the decline in global economic activity, exports, and inbound tourist flows, the reduction in mobility and consumption, and the impact of uncertainty on investment by firms.

In 2021 with the rapid progress of the vaccination campaigns, global GDP growth has strengthened, and the outlook is improving, although with differences across countries. Monetary policy remains expansionary in all the major countries. In Italy, GDP growth was slightly positive in the first quarter, and the recovery was driven above all by investment.

On November 1, 2011, Draghi was appointed President of the European Central Bank. He was succeeded as Governor by Ignazio Visco, who is still in office, having been confirmed in 2017 for his second mandate.

In Visco's CF, the highest frequency of the term *crisis (crisi)* is recorded: it appears 44 times in 2016 and 25 times on average during his 10 years. The focus is, in fact, on the financial crisis and the reforms necessary to counter it; in the last two CF, focus shifts onto the crisis from the COVID-19 *pandemic (pandemia)*.[3]

Furthermore, there are more references to *technologies (tecnologie)*, increasingly fundamental with the spread of the digital economy and to be able to offer services with greater added value.

For the first time, the term *non-performing loans (crediti deteriorati)* is one of the top ones. The use of *growth (crescita)* and *market (mercato)*, together with *risk/s (rischio/i)* and *investment/s (investimento)* increased. In Visco's CF, there is great attention to the Eurosystem's actions aimed at countering the sovereign debt crisis and deflation risks: *monetary policy (politica monetaria)* is frequently used as well. Also, there is a constant care to human capital, youth, school, university, and educational issues and new items appear, such as customer protection, financial education, and topics concerning civil justice and litigation. For the first time, in

[3] In the past, local peaks in the use of the keyword "crisis", correspond to the Korean War (1952), the Suez crisis, the oil crises of the seventies and those of the nineties (Argentina, Asian countries, Russia, Long-Term capital management). However, the systematic use of this keyword has exploded since 2008.

Visco's CF, topics concerning *Fintech* are outlined in his last four texts, and themes related to ***environment (ambiente/sostenibilità),*** the so-called ESG factors, are described in his last two CF (2019 and 2020).

8.5.3 *Keywords*

In this Section, we discuss the distribution of 50 *keywords* whose meaning is strictly connected with the contents discussed in Sect. 8.3.2. Most such items are technical terms, while others belong to common lexicon (such as for instance *"paese"—"country"* or *"famiglia"—"family")* which are, yet, used by the Governors to refer to specialized contents.

Table 8.6 shows the ranking of the keywords in the 75 CF. Six (*paese, mercato, credito, imprese, prezzo, investimento*) appear in all the CF. Two (*sviluppo* and *capitale*) occur in all but one CF.

The distribution of the keywords shows that:

(i) there is a clear link between the lexical choices of each Governor and the events that have happened to the Italian economy across the decades;
(ii) in line with our expectations, frequency of occurrence is variable across the years, thus paralleling the degree of attention given to each specific economic/financial topic at each given historical time-interval;

Figure 8.2 shows the correlation between *"investimento/i" (investment/s)* and *"imprese" (enterprises),* which is particularly high during the most recent governorships. Figure 8.3 shows the correlation of *"crisi" (crisis)* and *"rischio/i" (risk/s),* which increased especially after the 2000s, during Draghi's and Visco's mandates. "Rischio/i" is Draghi's second most used keyword (20 times per document), immediately followed by "crisi" (19). Visco uses "crisi" 25 times and "Rischio" 24 times. Figure 8.4 shows the correlation between *"inflazione" (inflation)* and *"cambio" (exchange),* which occur with greater incidence during critical moments.

First-time adoption of each keyword (Table 8.7) is linked to the vision of each Governor, as well as to contingent events. The items *aspettative (expectations)* and *innovazione (innovation)* were first used by Menichella in 1951 and 1954, respectively. Ten years later (1964), Carli used for the

Table 8.6 Keywords

Keywords(ITA)	*Keywords (EN)*	*Total*	*# of 0 per CF*	*Keywords (ITA)*	*Keywords (EN)*	*Total*	*# of 0 per CF*
Paese/i	Country/ies	3.370	0	Sistema bancario	Banking system	418	4
Mercato	Market	2.224	0	Famiglie	Households	394	19
Credito	Credit	2.054	0	Prodotto interno lordo/PIL	Gross domestic product/GDP	374	27
Imprese	Firms	1.877	0	Settore pubblico	Public sector	295	22
Prezzo/i	Price/s	1.854	0	Aspettative	Expectations	219	17
Investimento/i	Investment/s	1.724	0	Innovazione/i	Innovation/s	216	14
Sviluppo	Development	1.542	1	Sistema creditizio	Credit system	213	26
Capitale/i	Capital/s	1.360	1	Tecnologia/e	Technology/ies	210	25
Crescita	Growth	1.121	21	Materie prime	Raw materials	208	25
Lavoro	Labour	1.070	2	Disoccupazione	Unemployment	205	10
Cambio/i	Exchange/s	1.009	5	Debito pubblico	Public debt	203	22
Inflazione	Inflation	899	5	Area dell'euro	Euro-area	173	52
Liquidità	Liquidity	890	3	Borsa	Stock exchange	126	32
Produzione	Production	886	2	Tecnologico/a/i/che	Technological	115	35
Risparmio	Saving/s	841	2	Stabilità monetaria	Monetary stability	114	31
Crisi	Crisis	793	6	Sofferenze	Bad exposures/Bad loans	68	46
Rischio/i	Risk/s	778	7	Deteriorati	Non performing loans/NPL	58	65
Espansione	Expansion	772	2	Economie emergenti	Emerging economies	48	54
Moneta	Money	665	5	Pandemia/pandemico	Pandemia/pandemic	45	73
Riserve	Reserves	618	10	Derivato/i	Derivative/s	34	61
Aziende di credito	Banks (before 1993)	567	29	Globalizzazione	Globalisation	21	60

(continued)

Table 8.6 (continued)

Keywords(ITA)	*Keywords (EN)*	*Total*	*# of 0 per CF*	*Keywords (ITA)*	*Keywords (EN)*	*Total*	*# of 0 per CF*
Politica monetaria	Monetary policy	491	6	Cartolarizzazione/i	Securitisation	16	66
Vigilanza	Supervision	482	6	Sostenibilità/Sostenibili/e/ESG	Sustainability/sustainable/s/ESG	14	73
Tassi di interesse	Interest rates	449	14	Fintech	Fintech	7	71
Concorrenza	Competition	437	4	Incagli/Inadempie nze probabili	Unlikely to pay	6	69

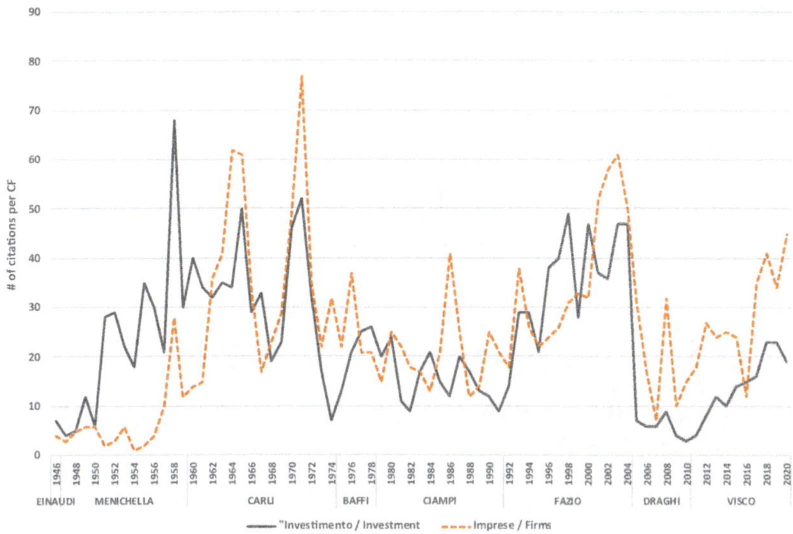

Fig. 8.2 Investimento/i vs. Imprese (*Investment/s vs Firms*)

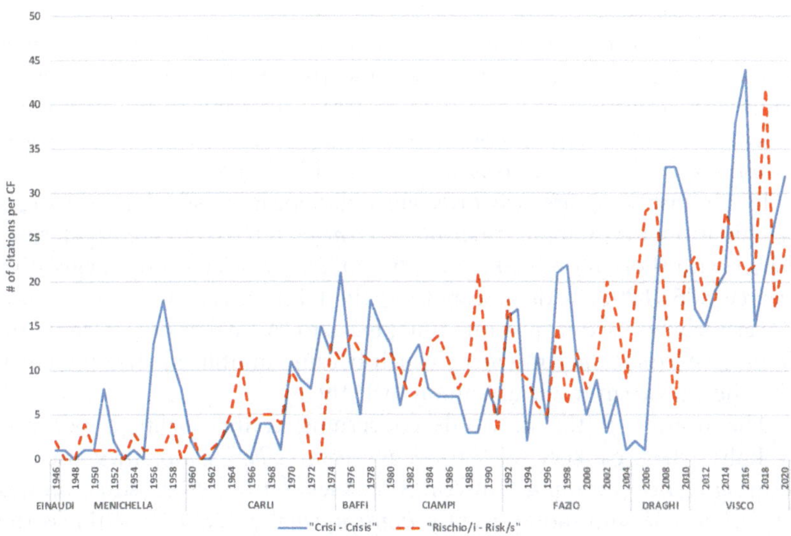

Fig. 8.3 Crisi vs. Rischio/i (*Crisis vs Risk/s*)

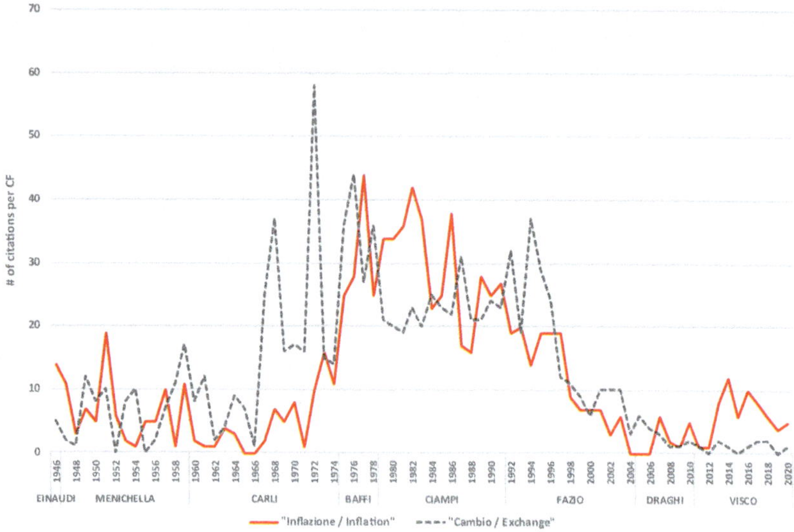

Fig. 8.4 Inflazione vs. Cambio (*Inflation vs Exchange*)

first time *tecnologia (technology)*. The item *derivati (derivatives)* was first used in 1992, to refer to potential systemic risks, while *economie emergenti* (emerging countries) was first used in 1993, referring to the very rapid growth pace of emerging countries as compared to industrialized ones; *globalizzazione (globalization)* was first used in 1996, by Fazio.

In contrast, there are keywords which disappear, mostly due to changes in legislation; for example, *"aziende di credito"* referred to credit intermediaries with short-term lending and funding activities until the Legislative Decree 385/1993. It disappears from the CF precisely in 1993: the last Governor who used it, just one time, is Fazio. A further case is *"stabilità monetaria" (monetary stability)*, which became unutilized after the entry into the European monetary union in 1999.

The history of the keywords concerning credit quality is peculiar: in Italy, when bankers refer to *bad exposures/bad loans,* they use the term *sofferenze* (meaning "sufferings"). A less serious credit situation was labeled *incagli* and more recently *inadempienze probabili* and this is the overall *unlikely to pay* category. The use of these keywords follows the national quality credit history: the first Governor who mentioned them was Baffi in 1975 and 1976.

Table 8.7 Keyword First time adoption

Keywords (ITA)	*Keywords (EN)*	*# of CF in which the keyword is missing at all*	*First time adoption*	*Governor*
Aspettative	Expectations	17	1951	Menichella
Innovazione/i	Innovation/s	14	1954	Menichella
Settore pubblico	Public sector	22	1956	Menichella
Crescita	Growth	21	1957	Menichella
Tecnologia/e	Technology/ies	25	1964	Carli
Tecnologico/a/i/che	Technological	35	1964	Carli
Sofferenze	Bad exposures/Bad loans	46	1975	Baffi
Incagli/Inadempienze probabili	Unlikely to pay	69	1976	Baffi
Derivato/i	Derivative/s	61	1992	Fazio
Economie emergenti	Emerging economies	54	1993	Fazio
Globalizzazione	Globalisation	60	1996	Fazio
Area dell'euro	Euro-area	52	1998	Fazio
Cartolarizzazione/i	Securitisation	66	2001	Fazio
Deteriorati/npl	Non performing loans/NPL	65	2011	Visco
FinTech	FinTech	71	2017	Visco
Pandemia/pandemico	Pandemia/pandemic	73	2019	Visco
Sostenibilità/sostenibile/i/ESG	Sustainability/sustainable/s/ESG	73	2019	Visco

8.6 Conclusions

This work proposed a preliminary attempt to analyze specialized economic/financial texts through an interdisciplinary lens, using a combination of competences from economics/finance and linguistics.

The choice of the CF was motivated by their very nature of official technical documents, published by a national institution, and addressed to a specialized audience. As such, they linguistically belong to a peculiar textual typology that resembles scientific texts but also technical reports and, to a lesser extent, public speeches. The linguistic features of the CF that emerged from our analysis can be summarized as follows:

(a) They are highly specialized texts: as such, they adopt a variety of language that coincides with the specialized variety of economy and finance.
(b) Changes in lexical choices depend on the content of each document and on more general trends in word-use observed in the Italian language over the years.
(c) Yet, the core structural features seem not to have undergone any significant change.

Some further, more general conclusions also emerged from our work, which suggest that a "multifocal" approach to the investigation of economic texts may provide viewpoints which would not be reached through a monodisciplinary lens. Our results can be summarized as follows:

(1) The analysis of the contents of the CF allowed us to sketch a very precise overview of the economic and financial history of the past century in Italy, which is summarized in Sect. 8.3.2.
(2) The observation of selected lexical choices (8.3.2 and 8.3.3) allowed us to highlight interesting correlations between these lexical choices and specific historical trends, and between different types of contents, which had never been highlighted before. As expected, the correlation between contents and lexical choices is quite high: the two components are non-separable.
(3) Technical documents must meet the requirement of a precise and detailed transmission of contents, which can only be met through rigorous linguistic choices. To achieve these goals, the scientific

community uses specialized varieties, which, yet, are often unintelligible to non-specialized audiences. The latter, though, especially in some fields of economy and finance, are interested in understanding at least some of the specialized contents of the discipline, and therefore must be put in the condition of understanding them even in the absence of a specialized knowledge. This, besides bringing up ethical and educational issues concerning financial alphabetization, raises the issue of "translating" specialized texts into a non-specialized variety, and adapting their linguistic shape to the linguistic background of non-expert speakers. This is a non-trivial issue, because translations of specialized texts into non-specialized varieties involve radical transformations of the original texts, often with a high amount of information loss.

We believe that, once refined and better intermingled, a multidisciplinary approach to economic texts along the lines we preliminarily proposed in this paper can open novel research lines, especially in the field of economic and financial communication.

References

Alberici, A. (2006). *Aspetti organizzativi di Banca d'Italia e del sistema nelle Considerazioni finali del Governatore*, fb–banca in formazione, quarterly of finance&banking–Associazione per lo sviluppo organizzativo e delle risorse umane, 42: 5–22. http://wp.demm.unimi.it/files/wp/2007/DEMM-2007_026wp.pdf

Astuti, V., De Bonis, R., Marroni, S., & Vinci, A. (2020). *Così parlarono i Governatori della Banca d'Italia: un'analisi del corpus linguistico delle Considerazioni finali.* Questioni di Economia e Finanza (Occasional Papers), Banca d'Italia, 592. https://www.bancaditalia.it/pubblicazioni/qef/2020-0592/QEF_592_20.pdf

Barbiellini Amidei, F. (2020). Paolo Baffi, una lezione civile. *Bancaria, 4*, 2–9.

Barucci, P. (2008). *Luigi Einaudi, Considerazioni Finali della Banca d'Italia.* Treves editore.

Berruto, G. (1995). *Fondamenti di sociolinguistica.* Laterza.

Bianchi, B., De Rosa, L., & Toniolo, G. (1998). Sull'opera di Donato Menichella. *Rivista Di Storia Economia, 1*, 63–84.

Bruno, G. (2016). *Text mining and sentiment extraction in central bank documents.* https://doi.org/10.1109/BigData.2016.7840784. https://ieeexplore.ieee.org/document/7840784

Chambers, J. K. (2003). *Sociolinguistic theory*. Blackwell.
Ciocca, P. (2018a). Carlo Azeglio Ciampi: Un profilo da "via Nazionale." *Bancaria, 1*, 2–9.
Ciocca, P. (2018b). Ciampi, Prodi e la lira a quota 990. *Bancaria, 5*, 16–17.
Cortelazzo, M. (1994). *Lingue speciali. La dimensione verticale*. Unipress.
Dardano, M. (1998). Il linguaggio dell'economia e della finanza. In I. Domenighetti (Ed.), *Con felice esattezza. Economia e diritto tra lingua e letteratura* (pp. 65–88). Casagrande.
Edwards, J. (1994). *Multilingualism*. Routledge.
Fazio, A. (1993, novembre 11). *Convegno in ricordo di Guido Carli*. BNL, Associazione Bancaria Italiana. https://www.bancaditalia.it/pubblicazioni/interventi-governatore/integov1993/Fazio_111193.pdf
Fazio, A. (2003, aprile 23). *Guido Carli Governatore della Banca d'Italia*. LUISS, Associazione Guido Carli. https://www.bancaditalia.it/pubblicazioni/interventi-governatore/integov2003/fazio_23_04_03.pdf
Gai, L. (1995). *Le Considerazioni Finali come strumento di moral suasion. La loro incidenza sul contesto normativo ed operativo dell'attività creditizia*. Giappichelli.
Gigliobianco, A. (2019). Guido Carli: Finanziare gli investimenti. *Bancaria, 1*, 81–88.
Gigliobianco, A., & Massaro, R. (2010). Paolo Baffi dietro le quinte. *Bancaria, 9*, 76–86.
Gotti, M. (1991). *I linguaggi specialistici*. La Nuova Italia.
Hymes D. H. (1966). *On communicative competence*. University of Pennsylvania Press, 1971
Labov, W. (1972). *Sociolinguistic patterns*. University of Pennsylvania Press.
Labov, W. (2001). *Principles of linguistic change. Social factors*. Wiley Blackwell.
Ligabue, V. (2021). *Considerazioni Finali della Banca d'Italia: contenuti economico-finanziari e scelte linguistiche*. MA Dissertation. Università di Modena e Reggio Emilia
Patuelli, A. (2016a). Donato Menichella: Doti tecniche e fiuto politico. *Bancaria, 4*, 104–106.
Patuelli, A. (2016b). Luigi Einaudi: Rigore etico e chiarezza di linguaggio. *Bancaria, 12*, 70–72.
Peluffo, P. (2007). *Carlo Azeglio Ciampi—L'uomo e il presidente*. Rizzoli.
Proietti, D. (2010). Economia, lingua dell'. In *Enciclopedia dell'italiano*. https://www.treccani.it/enciclopedia/lingua-dell-economia_(Enciclopedia-dell%27Italiano)
Rainer, F. (2006). Geschichte der Sprache der Wirtschaft in der Romania. In G. Ernst et al. (Eds.), *Romanische Sprachgeschichte. Ein internationales Handbuch zur Geschichte der romanischen Sprachen* (pp. 2148–2161). de Gruyter. Vol. 2.

Rando, G. (1990). "Capital gain", lunedì nero, "money manager" e altri anglicismi recentissimi del linguaggio economico borsistico-commerciale. *Lingua Nostra, 51*, 50–66.

Sarcinelli, M. (1999). La tutela del risparmio nel pensiero e nell'azione del Governatore Baffi. *Moneta e Credito, 206*, 225–240.

Sarcinelli, M. (2013). Le "Considerazioni finali" di Paolo Baffi. *Bancaria, 11*, 15–25.

Savona, P. (2008). *Sulle Considerazioni finali di Luigi Einaudi, testo dell'intervento effettuato l'11 dicembre 2008 in occasione della pubblicazione Luigi Einaudi*—Considerazioni Finali della Banca d'Italia, a cura di Piero Barucci. Treves editore.

Savona, P. (2013). Il modello economico di Paolo Baffi. *Bancaria, 11*, 26–33.

Saussure, F. (1916). *Cours de Linguistique générale*. Payot.

Signorini, L. F. (2021, ottobre 20). *Le "considerazioni finali" di Luigi Einaudi*. https://www.bancaditalia.it/pubblicazioni/interventi-direttorio/int-dir-2021/Signorini_IULM_20_ottobre_2021.pdf

Valente, A. (1990). *Da Einaudi a Ciampi. Le considerazioni finali dei governatori della Banca d'Italia*. 1947/1986. Laterza.

CHAPTER 9

Drivers of Shareholder Value Creation in M&A: Event Study of the European Banking Sector in the Post-financial Crisis Era

Gimede Gigante, Mario Baldacchini, and Andrea Cerri

9.1 Introduction

The creation of shareholder value for companies involved in M&A transactions is one of the most discussed topics in the academic literature. Many authors have questioned whether an organization's choice to grow by external means benefits shareholders or destroys value. In this context, a series of studies have focused on operations characterizing the banking sector, often obtaining conflicting results. Few studies, however, have

G. Gigante (✉)
Department of Finance, Bocconi University, Milan, Italy
e-mail: gimede.gigante@unibocconi.it

M. Baldacchini · A. Cerri
Bocconi University, Milan, Italy

P. Wachtel et al. (eds.), *Creating Value and Improving Financial Performance*, Palgrave Macmillan Studies in Banking and Financial Institutions, https://doi.org/10.1007/978-3-031-24876-4_9

analyzed the European market: until the 1990s, lack of cohesion among countries and of unified regulation made it difficult to identify common determinants of value creation. In the early 2000s, progressive integration at the European supranational level opened up new horizons and resulted in a growing body of related scholarship. Yet more recently, the 2008 financial crisis completely upset the traditional competitive logic of the sector, instituting radical and irreversible changes. New parameters have consequently emerged for evaluating the ability of banks to generate value for *shareholders*. The main objective of this chapter is therefore to identify the new drivers guiding the reactions of market participants to announcement of an acquisition. While the literature predating the financial crisis often focused on characteristics common to acquiring companies that could explain abnormal returns occurring on the date of *disclosure* of a corporate finance transaction, this study instead pays particular attention to the characteristics of target companies. The underlying hypothesis is that the market, in a period of crisis, awards premium returns to acquiring companies if the acquired credit institution has high levels of operational efficiency, a good ability to manage impaired positions, and sound capitalization.

The rest of the chapter is organized as follows. In Sect. 9.2, we recapitulate the main drivers of the consolidation of the European banking market in the period after the global financial crisis. We then summarize the pertinent background literature (Sect. 9.3). Next, Sect. 9.4 delves into our event study methodology, specifies the testable hypotheses and defines the sample selection. In Sect. 9.5, we present the main results of our analysis. Finally, Sect. 9.6 concludes reviewing the main thrust of our results, ventilate the potential limitations of our study, and sketches possible future avenues of research.

9.2 The European Banking Sector in the Post-Financial Crisis Era

The need to strengthen the European banking system in the years following the financial and sovereign debt crisis resulted in a series of both sector-specific and macroeconomic structural interventions. In this regard, three reforms have strongly impacted the operation and profitability of credit intermediation since 2010: (i) European Central Bank (ECB) monetary policy and interest rates, (ii) Non-Performing Loans and IFRS 9, and (iii) new capital requirements. Though these reforms have had negative consequences on growth outlook in the banking industry, they have increased the solidity of the sector and provided greater protections for the savings of account holders.

The highly expansive ECB monetary policy (i), implemented through acquisitions of assets on the open market (*Asset Purchase Program*) and *Long-Term Refinancing Operations*, has brought the main reference rates into negative territory. While on the one hand, this has allowed for an increased supply of credit to the real economy, enabling economic recovery, on the other hand, the drop in interest rates has also influenced the yields of loans, putting pressure on the Net Interest Margin, the main revenue source for banks.

Another factor weakening bank performance has been the explosion of Non-Performing Loans (ii), triggered by the economic crisis and poor growth prospects in the Eurozone. Recognizing the seriousness of the situation, banking authorities have proposed a reform of the accounting standards used by financial instruments so that provisions made against impaired positions better reflect expected and not incurred losses. The resulting IFRS 9 went into effect in 2014—replacing IAS 39, which was deemed inadequate for preventing situations of financial stress. Implementation of this new standard and the addenda introduced by the ECB, however, have had negative consequences on the profitability of banks: according to a 2019 PWC study, the initial adoption of IFRS 9 cost banks an average increase of 9% in loan loss provisions and about 51 basis points in terms of Core Tier 1 ratios.

Finally, the tightening of regulations regarding capital requirements (iii) has limited the ability of financial institutions to exploit the leverage effect and distribute wealth to shareholders in order to improve their resilience to high-stress scenarios. The new European CRR II and CRD V directives have further raised minimum CET1 ratios for *Global-Systemically Important Banks* to reduce systemic risk.

This period of crisis for the European banking sector has also had consequences on M&A activity. Since 2010, there has been a steady decline in both M&A volumes and values compared to the previous decade. Interestingly, in the last 20 years Italy has far outperformed other European countries in terms of the number of acquisitions, claiming about 26% of the total number of European transactions effected since 2000. There has also been a significant reduction in cross-border deals, which have literally stopped in the last decade. Analysis of deals closed in the last 20 years moreover reveals a clear preference for the acquisition of unlisted targets.

9.3 Literature Review

The literature on mergers and acquisitions in the banking sector has been heavily influenced by the evolution of the global financial system. In the 1980s and 1990s, the deregulation of lending in Europe and the U.S. gave rise to a *merger wave* that was also supported by increased globalization and shrinking cultural distances among countries. Despite a brief pause in 2002–2003, this consolidation process was interrupted less by the global financial crisis of 2007–2008 than by the consequences triggered by that crisis.

M&A was a key theme in the literature of the pre-crisis years, fueling several strands of research. One strand focused on identifying why banks seek external sources of growth (Focarelli & Pozzolo, 2001; Focarelli et al., 1999; Hagendorff et al., 2008; Hernando et al., 2008; Pasiouras et al., 2007). Among the key findings in this path of inquiry was that diversification, in terms of both product (e.g., bancassurance agreements) and geographical area, influences the likelihood that a credit institution will undertake an acquisition. The attractiveness of a given country's banking sector, possibility of exploiting economies of scale, and potential advantages resulting from the restructuring of underperforming companies have also been found to affect acquisition processes. Another strand of literature much debated in the first decade of the 2000s focused on the merger premium. On this topic, American literature has been more prolific than the European scholarship, whose main findings are encapsulated in two studies: Diaz and Azofra (2009) and Hagendorff et al. (2010). These studies find that acquiring companies seem willing to pay more for targets with high growth rates and lower risk if located in countries with less stringent regulation.

Though M&A has therefore taken a back seat in recent year compared to more topical issues, such as the impact of regulation or the determinants of bank performance, the creation of shareholder value has always remained central to the scholarship. This primary position derives from the multiple applications of knowledge about shareholder value creation: in addition to indicating whether shareholders and investors benefit on the day a transaction is announced, research on the creation of shareholder value also enables identification of the determinants influencing increases in stock market returns. Tourani and Van Beek (1999) were among the first to apply the methodology of *short-term event studies* in the European market, finding negative but not significant announcement returns for the bidding company. On the other hand, evidence of positive and statistically significant returns has been found in research on target companies. Using a sample of 54 *deals* selected from between 1988 and 1997, Cybo-Ottone and Murgia (2000) found statistically significant positive returns at the announcement date for the shareholders of bidding companies. The authors also ascertained that the creation of value was greater in cases in which deals were directed toward banks within the same country or toward financial institutions that would enable greater product diversification for the acquiring company. In the wake of these results, Lepetit et al. (2004) also found evidence of non-significant positive returns for acquiring companies, particularly when the transaction is made with the purpose of product diversification (e.g., credit institutions in insurance).

According to a study by Campa and Hernando (2006), though returns for target companies are positive at announcement, returns for acquiring shareholders are significantly negative, even over longer periods of analysis (i.e., one month after the event). These authors find that, in the long run, both acquiring and target companies experience negative *abnormal returns*, even if they are not statistically significant. Hagendorff et al. (2008) analyzed the determinants of *abnormal returns* in detail for the first time, examining whether *investor protection* in a target's country (i.e., the level of protection enjoyed by the company's shareholders) influences how investors react to the announcement of a transaction. Performing a comparative analysis across Europe and the U.S., this study identified an inverse relationship of returns: acquiring banks realize higher returns when the target is located in an economy with a low level of protection (such as European economies). Analysis of the study's sample of 53 European mergers from 1996 to 2004 moreover confirmed the presence of statistically significant positive returns for acquiring shareholders.

After publication of Hagendorff et al. (2008), analysis of the determinants of *abnormal returns* became the central object of study in Asimakopoulos and Athanasoglou (2013), Beltratti and Paladino (2013), and all subsequent studies. Asimakopoulos and Athanasoglou (2013) found that acquiring shareholders benefit more from domestic transactions and from transactions between listed banks, concluding that acquisitions of smaller, less efficient intermediaries do not generate increases in shareholder wealth. Beltratti and Paladino (2013) shifted to the years of the financial crisis: the study focuses on value creation in M&A transactions from between 2007 and 2010, confirming the hypothesis that, due to uncertainty, investors react only partially at the announcement of a transaction but positively at its completion date. This finding contradicts the previous evidence of long-term negative returns (e.g., Campa & Hernando, 2006). According to Beltratti and Paladino (2013), returns at the announcement date are positively related to the ROE and leverage of the bidding company. It is thus likely that, during the crisis, market participants concluded that banks with more profitability and capitalization were better able to exploit synergies from an acquisition.

The impact of the financial crisis was also the central theme in Rao-Nicholson and Salaber (2015). Focusing on cross-border transactions, this study noted that, after 2007, only acquisitions involving buyers in developing countries and targets in developed countries generated positive and significant returns for shareholders. Kyriazopoulos and Drymbetas (2015) returned focus to the pre-crisis period and confirmed the absence of significant returns to bidding shareholders at the announcement of ***domestic deals***. However, taking a longer time period (−10, + 10 days) into account, ***abnormal returns*** became negative. The authors also concluded that a more balanced capital structure in the target company positively effects value creation. According to Kyriazopoulos (2016), in M&A transactions between banks in Eastern Europe, acquirers have positive and statistically significant returns if an acquisition is paid for with cash on hand and if the target's country is characterized by high industry competitiveness.

The most recent study in this area was conducted by Leledakis and Pyrgiotakis (2019). Analyzing returns for acquirers using a sample of 312 extraordinary transactions among commercial banks announced between 1998 and 2016, this study observed negative overall returns. However, after dividing the sample into "pre-crisis" (M&A from 1998 to 2008) and "crisis" (M&A from 2008 on) subgroups, the authors noted negative and statistically significant ***abnormal returns*** in the "pre-crisis period" and positive returns in the "crisis" period. Moreover, as in Beltratti and Paladino (2013), returns turned out to be positively influenced by the leverage of the acquirer and degree of market concentration of the target country (measured by the Herfindahl–Hirschman Index and Concentration-5 Index).

Brief mention should also be made of the findings of studies focused specifically on the US market. Unlike the European literature, US scholarship has obtained statistical evidence that the shareholders of bidding companies do not benefit from increases in wealth from acquisitions

(DeLong, 2001; DeLong & DeYoung, 2007). Brewer and Jagtiani (2013) have also found evidence of a positive relationship between acquisition of the status of Too-Big-To-Fail by bidding credit institutions and those institutions' returns at the announcement of the transaction through which they attained that status.

9.4 Research Methodology

9.4.1 *Event Study*

To analyze the creation of shareholder value on the date of a merger announcement, an *event study* was performed. This widely employed methodology measures the impact of an exogenous event on the market price of a security in the short term. *Event studies* assume the efficiency of markets and rationality of operators, presuming that the information transmitted by an event will be immediately reflected in the prices of securities.

The informational content of an event is measured through comparison of *abnormal returns* with a *benchmark* return, also called a normal return, which is calculated using a model for estimating expected returns. If abnormal returns differ in a statistically significant way from normal returns in the time window in which an event takes place, it is possible to conclude that the information contained in the event has had an impact on the value of the companies under analysis.

This chapter followed the event methodology proposed by Campbell et al. (1997), which involves the following steps: (i) event identification; (ii) sample selection; (iii) identification of *abnormal returns*; (iv) estimation procedure; (v) significance testing; and (vi) interpretation and conclusion. For step (i), an *event window* of 21 days was used, which is in line what is suggested by the literature; this time period included the 10 days both before and after the *event date* (day 0) and an estimation window of 240 days, which is equal to a time span extending from 251 to 11 days before the merger announcement date (250 *trading days* correspond approximately to one calendar year). Therefore, $L1 = T1 - T0$ is defined as the duration of the *estimation window* and $L2 = T2 - T1$ as the duration of the *event window*. Step (ii) will be discussed more extensively below. Steps (iii) and (iv) consist in the estimation of *abnormal returns*, i.e., returns considered abnormal compared to a benchmark expected return. In this study, the *market model* was used to estimate benchmark returns. This is the most frequently used model in the literature, due to its predictive power and simplicity of application.

After estimating the market model parameters based on the returns in the *estimation window*, it was possible to calculate the *abnormal returns* (ARs) of each security in the *event window* as:

$$\widehat{AR}_{i,t} = R_{i,t} - \widehat{\alpha}_i + \widehat{\beta}_i R_{m,t} \tag{9.1}$$

where $R_{i,t}$ represents the return of the *i*-th security at time *t* in the *event window*. It is common practice to aggregate Ars around the *event date* to obtain the Cumulative Abnormal Returns (CARs):

$$CAR_i(\tau_1, \tau_2) = \sum_{t=\tau_1}^{\tau_2} AR_{i,t} \tag{9.2}$$

where $T_1 \leq \tau_1 \leq \tau_2 \leq T_2$. . The significance tests under (v) are designed to test for the existence of a causal relationship between an *abnormal return* and an event. Using a sample of securities, returns can be further aggregated by security. Given a number of securities equal to *N*, the *Average Abnormal Return* (AAR) is defined for period t and the *Cumulative Average Abnormal Returns* (CAARs) are determined.

$$AAR_t = \frac{1}{N} \sum_{i=1}^{N} AR_{i,t} \tag{9.3}$$

$$CAAR(\tau_1, \tau_2) = \frac{1}{N} \sum_{i=1}^{N} CAR_i(\tau_1, \tau_2) \tag{9.4}$$

The following hypothesis test was therefore constructed:

$$H_0 : AAR_t = 0 \qquad H_1 : AAR_t \neq 0$$

$$H_0 : CAAR = 0 \qquad H_1 : CAAR \neq 0$$

Rejecting the null hypothesis, it is possible to conclude that an event has a statistically significant impact on the market value of the securities included in the sample. AARs and CAARs have an important statistical property: under the null hypothesis H0, they normally distribute with a mean of zero and variance equal to the variance of the error term of the *market model*.

The tests performed on the AARs and CAARs in this study were both parametric and nonparametric, since using only parametric tests could lead to unreliable conclusions if the assumption regarding the normal distribution of AARs were incorrect. The parametric tests performed were the *cross-sectional t*-test, *portfolio method t*-test, and the test proposed in Patell (1976). While in the first two, it is possible to *over-reject* the null hypothesis due to event-induced volatility, Patell's test uses *Standardized Abnormal Returns* (Ars). The non-parametric tests employed were the Corrado (1989) Rank Test and the *Cumulative Rank Test* proposed in Hagnäs and Pynonnen (2014). The latter test is essentially an extension of the former, since in its original formulation, the Corrado Rank Test was not intended for CARs.

In addition to analysis of the data (vi) obtained, a *cross-sectional* multiple linear regression was also performed to search for the determinants of CARs.

9.4.2 Objectives of the Analysis

As highlighted above, *short-term studies* of the European banking system have obtained mixed results, particularly in determining the Ars of an acquiring firm. Leledakis and Pyrgiotakis (2019), for instance, suggest that investor expectations have changed in the post-financial crisis period: bidding shareholders benefit on average from statistically significant value increases. The first objective of this chapter is therefore to confirm this effect on a sample of M&A transactions announced from 2010 to 2020 using the event study methodology described above.

In addition to measuring the magnitude of *abnormal returns* in the sample and their statistical significance, the second objective of this chapter is to identify some factors that influence the CARs. For this type of analysis, it is common practice to use a *cross-sectional regression* that takes the CAR of a given acquisition as the dependent variable with respect to some independent variables. European scholarship has devoted particular attention to analyzing the characteristics of the acquirer[1] and the operation,[2] underestimating the impact a target company can also

[1] Leledakis and Pyrgiotakis (2019), Kyriazopoulos and Drymbetas (2015), Beltratti and Paladino (2013).

[2] Hagendorff et al. (2008); Campa and Hernando (2006).

have on the bidding company's returns—not just in terms of profitability[3] but also in terms of operational efficiency, asset quality, level of capitalization, and diversification. This study therefore addresses a set of characteristics of target companies to evaluate if they impact the expectations of the acquirer. In the *cross-sectional* regression, the following assumptions were made:

Hypothesis 1 Asset Quality and Capitalization: investors should react positively to the acquisition of targets that have good asset quality and a good level of capitalization.

Hypothesis 2 Business Model: investors should react positively when the acquisition involves targets whose business models are efficient and have balanced exposure to interest rates.

Hypothesis 3 Hypothesis 3—Diversification: investors should react positively to deals that allow for diversification of the acquirer's business model.[4]

Hypothesis 4 Behavioral Finance and Merger Waves: according to the theory of behavioral finance and merger waves, on the announcement date of an extraordinary transaction, overvalued companies should experience negative returns.[5]

The first three hypotheses refer to variables exclusively linked to the characteristics of the target. The fourth hypothesis, on the other hand, verifies whether the theory of behavioral finance applies even in a period of crisis such as that experienced by the European financial sector in the last decade. In addition to these variables, other variables related to the bidding company have been found to have significant impact on CARs by some studies.

[3] Hagendorff et al. (2008) found a significant inverse relationship between the CAR and EPS of the target; Kyriazoupoulos (2016) found a non-significant positive relationship with ROE.

[4] DeYoung et al. (2009); Leledakis and Pyrgiotakis (2019), and Beltratti and Paladino (2013).

[5] John et al. (2013), Chidambaran et al. (2010), Rhodes-Kropf et al. (2005); and Shleifer and Vishny (2003).

9.4.3 Selection of the Sample

The *deals* sample was extracted from Securities Data Company (SDC) Platinum by the provider Refinitiv, with a reference time frame including all M&A deals announced from January 1st, 2010 to November 30th, 2020. The sector of each deal was identified by SDC's "DA" (i.e., Commercial Banks, Bank Holding Companies) and "DB" (i.e., Savings and Loans, Mutual Savings Banks) codes. The geographic area of the sample was Europe (understood in a broad sense to include neighboring countries that—due to trade relations, market affinity, and the presence of relevant international financial institutions—are considered part of the European banking system). In Table 9.1, you can find a summary of deals by country. To enable collection of the target variables, some minimum information about each deal was required for inclusion in the sample: transactions without a stated value for the acquisition or in which the percentage of acquired shares was less than 3% were excluded. Further adjustments were necessary to avoid overlapping *estimation windows* for acquisitions made by the same company, to eliminate NPL purchases by banking institutions operating in the *servicing* sector, and to merge transactions announced jointly for the same target but aimed at different groups of shareholders. The sample thus contained 153 transactions.

The necessity of adding the target variables to the *cross-sectional regression* also affected the sample size: target lending institutions whose data could not be found were removed from the cross-sectional analysis. The final sample for the multiple linear regression therefore included 108 transactions.

To allow for further layers of analysis, the sample was divided into sections. *Panels A* and *B* were identified based on the listing status of the target, with *Panel A* encompassing listed companies and *Panel B* unlisted companies. *Panels C*, *D*, and *E*, on the other hand, divided the sample based on the geographic macro-area of the acquirer: "Southern Europe," "Eastern Europe," and "Northwestern Europe," respectively.

Table 9.1 Summary of deals by country

Country	*Number of deals*
Italy	40
Spain	22
Russian Fed	15
Denmark	13
Poland	10
France	7
Switzerland	7
Greece	7
Norway	6
Austria	6
Turkey	5
Germany	4
Sweden	2
Slovenia	2
Romania	1
Georgia	1
Netherlands	1
Liechtenstein	1
Portugal	1
Bulgaria	1
Cyprus	1
	153

9.5 Empirical Results

9.5.1 Results of the Event Study

Tables 9.2 and 9.3 present the main results of the event study for the entire sample of 153 deals. The results show that bidding investors react positively to the announcement of a deal. The average return on day zero was +0.78% and was found to be statistically significant. In addition, the 3-day and 5-day CAARs around the *announcement date* were also positive and statistically significant, highlighting that *acquirers* achieve an average gain of +0.88% and +0.96%, respectively. Despite being significant under parametric tests, the 5-day CAARs did not pass the Rank Test. This suggests that either event-induced volatility or the assumption of a normal distribution of ARs might have led to erroneous rejection of the null hypothesis.

Analyzing event window returns over a longer time period, however, this observation was reversed: in the 10 days following the announcement, the CAARs of acquiring companies became negative, though not statistically significant. Specifically, the [0.5] and [0.10] CAARs were –0.02% and –0.51%. This could indicate that, after an initial positive reaction, the market enacts a downward revision of expectations regarding the value creation of companies involved in the transaction. Finally, analysis of CAAR [–10, +10], which covers the entire *event window*, demonstrates that acquirers experience statistically significant negative returns of –1.92%.

The gains experienced during the days around the announcement fail to offset the negative *abnormal returns* that occur when taking the longer time frame into consideration. These results are aligned with scholarship that has found evidence of positive, though not significant, CAARs (Cybo-Ottone & Murgia, 2000; Hagendorff et al., 2008; Lepetit et al., 2004). In addition, since the positive returns of the sample were also significant, the findings of the recent study by Leledakis and Pyrgiotakis (2019) seem confirmed by these results. It is therefore possible to conclude that changes in the banking sector in the last decade have also had an impact on the reactions of investors to announcements of M&A deals. Consolidation can be seen as a value-generating process, given the negative earnings outlook and high regulation of the banking sector. Finally, the negative CAR for the entire event window aligns with the similar non-significant results Kyriazopoulos and Drymbetas (2015) obtained using a panel of *domestic deals*.

Table 9.4 shows the results of the event study after division of the sample based on the listing status of the target into *Panel A* (listed) and *Panel B* (unlisted).

The CAARs obtained in Panel A are in line with the analysis of the entire sample: a positive effect can be observed around the time of announcement, but the overall effect is negative if the analysis is extended to include the entire *event window*. However, compared to the whole sample, the 3-day and 5-day CAARs, though positive, are not significant. Furthermore, the returns for CAAR [–10, 10] and for CAAR [0, 10] are decidedly negative, at –4.58% and –2.72%, respectively. Panel B, on the other hand, shows significant and positive returns in each *event window*. In particular, in the three days around the announcement date, bidders experience positive and significant ARs of +1.23%.

Joint analysis of the two panels reveals different behavior based on whether the acquiring company is involved in a transaction with a listed or unlisted credit institution. Though both cases demonstrate an increase in value at the *announcement date*, in cases where a target is unlisted, the acquirer's shareholders benefit from positive returns throughout the *event window*. Therefore, to test the hypothesis that the acquisition of an unlisted financial institution creates greater value for *acquiring* shareholders than the acquisition of a listed target, a *t*-test was performed to analyze the difference in averages between two independent samples. The results, reported in Table 9.5, seem to confirm the hypothesis, especially for CAAR [–10, 10] and CAAR [0, 10], whose *t*-test values are statistically significant at the 10% level.

Table 9.6 presents the results of the event study based on Panels C, D, and E, reflecting the following respective geographic areas of origin for acquiring companies: Southern Europe, Eastern Europe, and Northwestern Europe. Panel C, which represents Southern Europe, reveals considerable uncertainty: beyond the day of announcement, there is no statistical significance to confirm positive or negative returns on average in the identified time windows.

For *acquirers* in the region of Eastern Europe, M&A transactions do not create shareholder value. CAARs are negative in both the ten days leading up to a *deal* and the ten days following it, with an overall negative return in the event window of –4.63%. On the other hand, for *acquirers* in Northwest Europe, CAARs are positive: banking M&A deals create value. On the day of announcement, there is an average gain of 1.63%, which becomes 2.32% over a 3-day window and reaches 3.27% in a 5-day window. To test whether the difference between the two regions (Eastern and Northwest Europe) was significant, a *t*-test was also performed for the difference in averages of the two independent samples (Table 9.7). The results confirm that, on average, the shareholders of acquirers located in Northwestern European countries benefit from higher value creation than the shareholders of companies located in Eastern European countries.

There could be various reasons why reactions differ across the three geographic areas identified. The absence of statistical significance in the returns in Panel C could reflect the uncertainty characterizing the countries that suffered most from the sovereign debt crisis due to the fragility of their banking systems. The results for Panel D are partially in line with the findings of Leledakis and Pyrgiotakis (2016), which studied a sample of 69 M&A deals announced in Eastern European countries from 1995 to 2015.

9.5.2 Cross-Sectional Regression

This section presents the results of the *cross-sectional regression*, employing CAR [−1, +1] as the dependent variable with respect to the independent variables. The 3-day CAR was selected because of its statistical significance (demonstrated in Table 9.3). Table 9.8 displays the results of the multiple linear regression performed on the entire sample. Model (1) includes all variables, while model (2) includes only variables related to the acquired company; model (3) shows variables related to the acquirer and to the *deal*. This division aims to isolate the effect of the variables in relation to the target.

The variables related to the target company seem to have a greater influence on the 3-day CAR than the set of variables related to the type of deal and acquiring company: the Adjusted *R*-Squared coefficient of model (2) is higher than that of model (3).

The statistical significance of coefficients *T_NPL_Ratio* and *T_Equity_Loans* confirms *Hypothesis 1—Asset Quality and Capitalization*, according to which investors would prefer financial institutions with low NPL levels and high levels of capitalization. The variable *T_NPL_Ratio* assumes a negative quadratic relationship with returns at the announcement date. This could indicate that the market reacts positively to NPL ratio levels in the target that are judged sustainable and reacts negatively to excessive levels that could require costly de-risking in the future. The output of the model also presents results that are partially aligned with expectations for *Hypothesis 2—Target Structure*: on the one hand, the negative influence on the CARs of variables *T_Loans_Assets* and *T_Deposits_Assets* confirms that the market rewards acquisitions of target companies with more diversified balance sheet structures that are less linked to interest rates. On the other hand, the findings show no evidence for the hypothesis that the acquirer can create more value through acquisition of a credit institution with high operational efficiency, as measured through the variable *T_Cost_Income*. The coefficient associated with this variable is positive and statistically significant, suggesting that investors see a potential to exploit synergies by restructuring inefficient targets. *Hypothesis 3—Diversification* is not confirmed. In fact, in the output of

Table 9.2 Cumulative average abnormal return: entire sample

Entire sample (N = 153)	*(%)*	*Cross-sectional T-test T-value*	*Portfolio method T-test T-value*	*Patell test Z-value*	*Rank test Z-value*
−10	−0.09	−0.51	−0.49	−1.17	−0.85
−9	−0.28	−1.50	−1.44	−0.71	−0.99
−8	−0.02	−0.08	−0.08	−0.31	−0.61
−7	0.22	1.18	1.13	0.70	0.59
−6	−0.26	−1.42	−1.37	−1.46	−1.40
−5	−0.19	−1.02	−0.98	−0.53	−1.11
−4	−0.29	−1.56	−1.50	−1.55	−1.05
−3	−0.25	−1.34	−1.29	−0.95	−1.21
−2	0.47	2.53**	2.43**	2.25**	2.14**
−1	−0.22	−1.17	−1.13	−1.06	−1.04
0	**0.78*****	**4.22*****	**4.05*****	**4.14*****	**2.61*****
1	0.29	1.56	1.50	2.26**	1.09
2	−0.37	−1.97*	−1.89*	−3.04***	−2.34**
3	−0.01	−0.08	−0.08	−0.76	−0.90
4	−0.62	−3.33***	−3.20***	−3.06***	−0.62
5	−0.09	−0.51	−0.49	0.05	0.87
6	−0.46	−2.50**	−2.40**	−2.50**	−1.57
7	−0.15	−0.79	−0.76	−0.81	−0.29
8	−0.08	−0.42	−0.40	−0.16	0.74
9	0.10	0.55	0.53	−0.28	0.50
10	0.10	0.52	0.50	−0.23	−0.71

Average Abnormal Returns resulting from the event study based on the entire sample of 153 deals announced between 2010 and 2020. The symbols denote the following levels of statistical significance of with a two-tailed *t*-test: (*) significant at 0.1; (**) significant at 0.05; (***) significant at 0.01

model (1), the variable *T_Diversif* is not significant, despite having a positive influence on the dependent variable.

Moreover, the ***dummy variable D_Domestic*** highlights that transactions between companies within the same nation on average benefit from a positive return of +1.84% compared to cross-border transactions. In model (3), the dummy is also found to have greater impact and statistical significance. With respect to *Hypothesis 4—Behavioral Finance and Merger Waves*, there is insufficient empirical evidence to confirm the hypothesis that overvalued companies finance acquisitions through share trading. The interaction between *A_Pbv* and *D_Stocks* was also tested, but lack of significance led to its exclusion. Finally, the characteristics of

Table 9.3 Cumulative average abnormal return: entire sample

Entire Sample (N = 153)	*(%)*	*Cross-sectional T-test T-value*	*Portfolio method T-test T-value*	*Patell test Z-value*	*Rank test Z-value*
CAAR [–10,10]	–1.42	–1.80*	–1.73*	–2.27**	–1.89*
CAAR [–10,5]	–0.93	–1.31	–1.44	–1.42	–1.23
CAAR [–5,5]	–0.50	–0.86	–1.14	–0.80	–0.50
CAAR [–1,1]	0.86	2.59**	6.59***	2.94***	2.25**
CAAR [–2,2]	0.96	2.27**	4.46***	1.93*	0.98
CAAR [0,5]	–0.02	–0.10	–0.18	–0.29	0.27
CAAR [0,10]	–0.51	–0.99	–1.31	–1.64	–0.32

Cumulative Average Abnormal Returns resulting from the event study based on the entire sample of 153 deals announced between 2010 and 2020. The symbols denote the following levels of statistical significance with a two-tailed *t*-test: (*) significant at 0.1; (**) significant at 0.05; (***) significant at 0.01

Table 9.4 Event study results: Panel A & Panel B

	(%)	*Cross-sectional T-test T-value*	*Portfolio method T-test T-value*	*Patell test Z-value*	*Rank test Z-value*
Panel A: Listed (N = 67)					
AR [0]	1.17	3.97***	4.03***	2.88***	1.57
CAAR [–10,10]	–4.58	–3.40***	–3.45***	–3.12***	–2.50**
CAAR [–10,5]	–2.77	–2.36**	–2.73***	–2.17**	–1.46
CAAR [–5,5]	–1.85	–1.90*	–2.66***	–1.66*	–0.84
CAAR [–1,1]	0.38	0.74	1.99**	0.59	0.69
CAAR [–2,2]	0.42	0.64	1.33	0.03	–0.18
CAAR [0,5]	–0.91	–1.26	–2.39**	–2.20**	–1.00
CAAR [0,10]	–2.72	–2.79***	–3.90***	–3.31***	–2.02**
Panel B: Non-listed (N = 86)					
AR [0]	0.49	2.03**	1.84*	2.97***	2.00**
CAAR [–10,10]	1.04	0.95	0.86	0.08	–0.10
CAAR [–10,5]	0.50	0.53	0.55	0.18	–0.26
CAAR [–5,5]	0.56	0.71	0.88	0.55	0.13
CAAR [–1,1]	1.23	2.97***	7.12***	3.59***	2.57**
CAAR [–2,2]	1.38	2.59**	4.80***	2.69***	1.57
CAAR [0,5]	0.67	1.15	1.94*	1.71*	1.25
CAAR [0,10]	1.21	1.52	1.91*	1.15	1.53

Cumulative Average Abnormal Returns resulting from the event study, divided according to the listing status of the target company. The symbols denote the following levels of statistical significance with a two-tailed *t*-test: (*) significant at 0.1; (**) significant at 0.05; (***) significant at 0.01

Table 9.5 *T*-test difference between means of two independent samples

Panel B vs Panel A	*Panel B*	*Panel A*	*Difference*	T-*value*
AR [0]	0.49	1.17	–0.68	0.80
CAAR [–10,10]	1.04	–4.58	5.62	1.79*
CAAR [–10,5]	0.50	–2.77	3.27	1.13
CAAR [–5,5]	0.56	–1.85	2.41	0.99
CAAR [–1,1]	1.23	0.38	0.85	0.68
CAAR [–2,2]	1.38	0.42	0.96	0.56
CAAR [0,5]	0.67	–0.91	1.58	1.09
CAAR [0,10]	1.21	–2.72	3.92	1.93*

Results of a two-tailed *t*-test for the difference in means between two independent samples with different variances. Symbols denote the following levels of statistical significance with a two-tailed test: (*) significant at 0.1; (**) significant at 0.05; (***) significant at 0.01.

acquiring companies show low levels of significance when compared to the characteristics of targets. CARs are positively affected by the acquirer's Roe (*A_Roe*) and the number of deposits divided by total assets (*A_Deposits_Assets*). The variable *A_Serial* turns out to be significant only in model (3): "serial" acquirers who made more than 4 transactions in the 2010–2020 period were rewarded by the market in terms of announcement returns.

The models presented here were subjected to statistical analysis of outliers, high leverage and influence, heteroscedasticity (Breush–Pagan test), and multicollinearity (*Variance Inflation Factor*). No critical issues were found. However, three observations were eliminated from the 108 originally identified because they simultaneously indicated outliers, high leverage, and influence. This elimination significantly improved the overall significance of the model.

Models (4) and (5) shown in Table 9.9 respectively refer to the multiple linear regression of CAR [–1, +1] and all independent variables for cases in which listed or unlisted targets were involved in the transaction.

If the *output* of the *event study* revealed noteworthy results, the *cross-sectional regression* performed on both panels did not bring to light any relevant findings with respect to the hypotheses. In model (4), analysis of Panel A (transactions concluded with listed banks) revealed that investors focus on the deposits to total assets ratio for both target and acquirer company. The only significant variables in model (5) were those related

Table 9.6 Event study results: Panel C, Panel D, and Panel E

	(%)	*Cross-sectional T-test T-value*	*Portfolio method T-test T-value*	*Patell test Z-value*	*Rank test Z-value*
Panel C: South (N = 72)					
AR [0]	0.61	2.01**	2.05**	2.40**	2.42**
CAAR [−10,10]	−0.88	−0.63	−0.65	−0.52	−0.15
CAAR [−10,5]	−0.92	−0.75	−0.88	−0.45	−0.22
CAAR [−5,5]	−1.23	−1.21	−1.72*	−0.89	−0.16
CAAR [−1,1]	0.30	0.57	1.54	1.21	1.85*
CAAR [−2,2]	0.23	0.34	0.71	0.32	0.75
CAAR [0,5]	0.00	0.00	−0.01	−0.12	0.45
CAAR [0,10]	0.03	0.03	0.04	−0.27	0.77
Panel D: East (N = 38)					
AR [0]	0.22	0.66	0.67	0.55	−0.17
CAAR [−10,10]	−4.63	−2.99***	−3.05***	−3.49***	−3.31***
CAAR [−10,5]	−3.48	−2.57**	−3.01***	−2.95***	−3.16***
CAAR [−5,5]	−1.81	−1.62	−2.28**	−1.97**	−2.01**
CAAR [−1,1]	−0.03	−0.05	−0.12	−0.13	−1.18
CAAR [−2,2]	−0.68	−0.90	−1.88*	−0.85	−1.53
CAAR [0,5]	−1.58	−1.92*	−3.66***	−2.49**	−2.03**
CAAR [0,10]	−2.74	−2.45**	−3.45***	−3.11***	−2.22**
Panel E: North-West (N = 43)					
AR [0]	1.63	5.56***	5.72***	4.67***	1.93*
CAAR [−10,10]	0.33	0.24	0.25	−0.01	0.01
CAAR [−10,5]	1.01	0.86	1.01	0.60	0.73
CAAR [−5,5]	1.82	1.87*	2.66***	1.65	1.23
CAAR [−1,1]	2.32	4.55***	12.40***	3.82***	2.45**
CAAR [−2,2]	3.27	4.98***	10.50***	3.57***	1.96*
CAAR [0,5]	0.90	1.24	2.40**	1.53	1.41
CAAR [0,10]	0.21	0.22	0.31	0.40	0.49

Cumulative Average Abnormal Returns resulting from the event study divided according to Panels C, D and E, which identify the geographic area of origin of the purchasing company. The symbols denote the following levels of statistical significance of a two-tailed *t*-test: (*) significant at 0.1; (**) significant at 0.05; (***) significant at 0.01

to *Hypothesis 1—Asset Quality*. However, the low overall significance of the model does not allow us to make inferences or draw conclusions based on the data presented in the output.

In the analysis of the *cross-sectional regression* for Panels C, D and E, on the other hand, a critical point emerged regarding the sample size: while Panel C includes 54 observations, constituting just over 50% of the total

Table 9.7 *T*-test for the difference in means of two independent samples

Panel E vs Panel D	*Panel E*	*Panel D*	*Difference*	*T-value*
AR [0]	1.63	0.22	1.41	1.13
CAAR [−10,10]	0.33	−4.63	4.96	1.52
CAAR [−10,5]	1.01	−3.48	4.49	1.69*
CAAR [−5,5]	1.82	−1.81	3.64	1.53
CAAR [−1,1]	2.32	−0.03	2.35	1.70*
CAAR [−2,2]	3.27	−0.68	3.95	2.09**
CAAR [0,5]	0.90	−1.58	2.48	1.20
CAAR [0,10]	0.21	−2.74	2.95	1.03

Results of a two-tailed *t*-test on the difference in means between two independent samples with different variances. Symbols denote the following levels of statistical significance of a two-tailed *t*-test: (*) significant at 0.1; (**) significant at 0.05; (***) significant at 0.01

sample, Panels D and E, taken individually, did not contain a sufficient number of M&As to obtain significant results. It was, therefore, necessary to merge these panels to achieve significant results.

To account for the diversity across geographic areas, dummy variable *D_West* was added, which takes a value of 1 if the observation is classified in Panel E (Northwest Europe) and 0 if belonging to Panel D (Eastern Europe). This variable was added due to the results of the analysis of the differences among clusters in Table 9.10 and shows that *acquirers* located in the Northwest region obtained a higher average gain at the announcement date than those in the Eastern region.

The output of model (6) again highlights the relevance of NPLs, especially if interpreted in light of the weakness of the banking sector in the countries of Southern Europe. Furthermore, it seems evident that the market rewards *domestic* transactions with an average gain of +5.196%. This result is consistent with the need repeatedly stressed by the ECB to initiate a consolidation process to ensure the solidity of the national financial systems of Southern Europe, which are strongly dominated by small savings and cooperative banks. Here, the *event window* returns are also positively influenced by the ROE and M&A track records of acquiring companies, meaning the market positively assesses the acquirer's ability to generate value and increase its size through numerous acquisitions.

Finally, model (7) does not present any particularly relevant insights. Though confirming that transactions in the North-West cluster benefit from an average gain of +0.8% with respect to those in the East, the

Table 9.8 Cross-sectional regression: entire sample

	(1) Full model		*(2) Target*		*(3) Deal and acquirer*	
Constant	*Coeff.*	*T-value*	*Coeff.*	*T-value*	*Coeff.*	*T-value*
T_Npl_Ratio	–5.282	–1.09	–1.307	–0.31	–6.430	–2.99***
(T_Npl_Ratio)^2	0.393	2.59**	0.444	3.37***		
T_Equity_Loans	–0.010	–2.42**	–0.011	–3.21***		
T_Deposits_Assets	0.067	1.72*	0.057	1.50		
T_Loans_Assets	–0.079	–2.84***	–0.080	–2.87***		
T_Cost_Income	–0.037	–0.91	-0.021	–0.57		
T_Diversif	0.063	2.71***	0.071	2.96***		
D_Domestic	0.004	0.14	0.010	0.35		
D_Stocks	1.837	1.61			2.899	2.62**
A_Pbv	0.025	0.02			–0.550	–0.45
A_Run_Up	–0.041	–1.57			–0.020	–0.73
A_Roe	–0.027	–1.56			–0.014	–0.76
(A_Roe)^2	0.084	1.44			0.018	0.29
A_Deposits_Assets	0.003	1.70*			0.003	1.78*
A_Serial	0.057	1.75*			0.076	2.23**
A_Relsize	1.824	1.59			2.664	2.29**
	–0.848	–0.76			–0.046	–0.04
N	105		105		105	
F-Statistic	3.44		4.62		2.18	
Adj. R^2	0.27		0.20		0.09	

Output of multiple linear regression with dependent variable CAR [–1, +1]. Symbols denote the following levels of statistical significance of a two-tailed *t*-test: (*) significant at 0.1; (**) significant at 0.05; (***) significant at 0.01

dummy variable *D_West* results in a non-significant coefficient. Also in this case, Hypothesis-1 seems to be verified, since the significance of the NPL ratio of the target is confirmed.

9.6 Conclusions

The empirical evidence shows that when acquisitions in the European banking sector are announced, shareholders of the acquiring company benefit from positive returns on average. While many authors have tried to verify whether there are characteristics common to acquiring companies that influence value creation in M&A transactions, little attention has been given to the traits of target companies. This study therefore shows

Table 9.9 Cross-sectional regression: Panel A and Panel B

Variables	*(4) Panel A—Listed*		*(5) Panel B—Unlisted*	
	Coeff	*T-value*	*Coeff*	*T-value*
Constant	−4.464	−0.45	−6.911	−1.08
T_Npl_Ratio	0.356	1.73*	0.465	1.78*
$(T\_Npl\_Ratio)^2$	−0.008	−1.44	−0.012	−1.78*
T_Equity_Loans	−0.101	−0.81	0.087	1.90*
T_Deposits_Assets	−0.195	−3.36**	−0.046	−1.19
T_Loans_Assets	−0.024	−0.28	−0.022	−0.42
T_Cost_Income	0.076	1.45	0.045	1.39
T_Diversif	0.096	1.23	0.008	0.18
D_Domestic	0.983	0.55	−0.055	−0.03
D_Stocks	1.147	0.55	−2.145	−1.22
A_Pbv	−0.015	−0.44	−0.051	−0.96
A_Run_Up	−0.019	−0.76	−0.015	−0.52
A_Roe	0.076	0.41	0.029	0.26
$(A\_Roe)^2$	0.002	0.38	0.001	0.32
A_Deposits_Assets	0.133	2.67**	0.086	1.63
A_Serial	1.603	1.04	2.454	1.12
A_Relsize	0.837	0.40	−1.586	−0.98
N	44		61	
F-Statistic	3.372		1.617	
Adj. R^2	0.46		0.14	

Output of multiple linear regression with dependent variable CAR [−1, +1] for Panel A and Panel B. Symbols denote the following levels of statistical significance of a two-tailed *t*-test: (*) significant at 0.1; (**) significant at 0.05; (***) significant at 0.01

that the characteristics of a target financial institution, which relate to the main issues briefly discussed in the first section of this chapter—interest rates, NPLs, capital requirements, and declining M&A volumes—are among the main determinants of a bidding company's abnormal returns.

Segmenting the sample into different panels also revealed noteworthy results. When targets are listed entities, the CARs of acquirers are negative, whereas if targets are private companies, returns are positive. This could indicate that market participants negatively discount the possibility of integrating two corporate cultures and two different modus operandi that are already well-established in the market. Analysis of the determinants, however, did not reveal particularly significant results in economic terms.

Table 9.10 Cross-sectional regression: Panels C, D, and E

	(6) Panel C "Southern Europe"		*(7) Panel D & E "Rest of Europe"*	
Variables	*Coeff*	*T-value*	*Coeff*	*T-value*
Constant	−22.315	−2.50**	−4.711	−0.61
T_Npl_Ratio	0.516	1.70*	0.427	1.84*
(T_Npl_Ratio)^2	−0.019	−2.31**	−0.011	−1.75*
T_Equity_Loans	0.122	1.65	0.065	1.27
T_Deposits_Assets	−0.013	−0.34	−0.123	−2.73**
T_Loans_Assets	0.008	0.12	0.003	0.96
T_Cost_Income	0.050	1.23	0.021	0.59
T_Diversif	0.045	0.73	0.047	0.33
D_Domestic	5.196	2.03**	0.771	0.64
D_Stocks	−1.764	−1.02	−2.414	0.21
A_Pbv	−2.487	−1.13	−0.008	0.83
A_Run_Up	0.008	0.33	−0.052	0.15
A_Roe	0.046	0.71	0.731	0.15
(A_Roe)^2	0.007	2.93***	−0.028	0.16
A_Deposits_Assets	0.188	3.11***	0.039	0.42
A_Serial	5.218	3.18***	−3.817	0.34
A_Relsize	0.808	0.84	−0.395	0.78
D_West			0.800	0.69
N	54		51	
F-Statistic	3.606		1.819	
Adj. R^2	0.44		0.22	

Output of multiple linear regression with dependent variable CAR [−1, +1] for Panels C, D, and E. Symbols denote the following levels of statistical significance of a two-tailed *t*-test: (*) significant at 0.1; (**) significant at 0.05; (***) significant at 0.01

On the other hand, division of the transactions by geographic area brought to light differences among the various regions of the Eurozone, especially with regard to Southern Europe. In fact, while there are positive CARs for acquirers in Northwestern Europe and negative CARs for those in Eastern Europe, returns in Southern Europe are more uncertain. This situation could be explained by the fragility of the financial systems in Southern European countries, which suffered most from the sovereign debt crisis and have faced drastic austerity policies to contain public debt. This hypothesis, which emerged through analysis of the results of the *event study*, could be confirmed by the significance of certain coefficients in the output of the model. Firstly, the attention paid by investors to asset quality. Secondly, the theme of domestic banking consolidation mentioned above, which was underscored by the +5.19% return that acquiring companies involved in domestic transactions gained on average. Finally, the characteristics of acquiring companies were found to be more significant than those pertaining to the full sample. Market participants are more confident about value creation in M&A transactions if acquirers have high levels of profitability and a solid track record of external growth.

Though the results obtained by this study demonstrate a good level of significance overall, the analysis has some limitations. The first is linked to the assumptions underwriting use of the *event study* methodology. Though choice of the *market model* as the method for estimating expected benchmark returns is recognized in the literature as the most effective, its reliability depends exclusively on the R-Squared level of the simple linear regression in relation to the market portfolio (Stoxx Europe 600 Banks index). Furthermore, use of the *market model* assumes the normality of daily returns, which could often be erroneous. To overcome these problems, multifactor models could be used, such as the three-factor Fama–French model; this would allow for more specification among the *abnormal returns* obtained. A second limitation of this study concerns the sample size for the *cross-sectional regression*. Although the full sample of 153 transactions is in line with much of the literature in terms of quantity, the reduction of the sample due to lack of data about some target companies could have deprived the analysis of some important observations. However, it is also worth noting that the absence of data mainly affected small transactions with sizes that were significantly below average. Such operations could hardly have influenced the distribution of CAARs or AARs. Finally, a larger sample size would have allowed for more meaningful results from the *cross-sectional regression* analysis performed on the panels divided by geographic area of origin.

This study's results offer numerous insights for future work in event studies and the European banking industry. Many analysts agree that it will be necessary in the next few years to resume the financial consolidation process interrupted in 2007, for reasons of both competitiveness and recovery of profitability. In fact, most credit institutions have already made all the capital adjustments necessary to ensure full compliance with discretionary requirements; avoiding having to recapitalize distressed institutions or to carry out further *de-risking* operations will make the merging of organizations easier. In conclusion, the onset of a new wave of M&A in the industry could once again shift the focus of the literature back toward finding new drivers to explain value creation. Along this, it could be interesting to address the potential role of ESG scores in shaping the M&As activities in banking. Indeed, as suggested by the previous three chapters, ESG scores synthesize a new view of value creation and it would be worth studying whether this drives bank M&As beyond shareholder value creation.

References

Asimakopoulos, I., & Athanasoglou, P. (2013). Revisiting the merger and acquisition performance of European banks. *International Review of Financial Analysis, 29*(C), 237–249.

Beltratti, A., & Paladino, G. (2013). Is M&A different during a crisis? Evidence from the European banking sector. *Journal of Banking and Finance, 37*, 5394–5405.

Brewer, E., & Jagtiani, P. (2013). How much did banks pay to become too-big-to-fail and to become systemically important? *Journal of Financial Services Research, 43*(1), 1–35.

Campa, J. M., & Hernando, I. (2006). M&As performance in the European financial industry. *Journal of Banking and Finance, 30*, 3367–3392.

Chidambaran, N. K., John, K., Shangguan, Z., & Vasudevan, G. (2010). Hot and cold merger markets. *Review of Quantitative Finance and Accounting, 34*(3), 327–349. https://doi.org/10.1007/s11156-009-0133-z

Corrado, C. J. (1989). A nonparametric test for abnormal security-price performance in event studies. *Journal of Financial Economics, 23*, 385–395.

Cybo-Ottone, A., & Murgia, M. (2000). Mergers and shareholder wealth in European banking. *Journal of Banking & Finance, 24*(6), 831–859.

DeLong, G., & DeYoung, R. (2007). Learning by observing: Information spillovers in the execution and valuation of commercial bank M&As. *Journal of Finance, 62*(1), 181–216.

DeLong, G. L. (2001). Stockholder gains from focusing versus diversifying bank mergers. *Journal of Financial Economics, 59*(2), 221–252.

DeYoung, R., Evanoff, D., & Molyneux, P. (2009). Mergers and acquisitions of financial institutions: A review of the post-2000 literature. *Journal of Financial Services Research, 36*(2), 87–110.

Diaz, B., & Azofra, S. S. (2009). Determinants of premiums paid in European banking mergers and acquisitions. *International Journal of Banking, Accounting and Finance, 1*(4), 358–380.

Focarelli, D., Panetta F., & Salleo C. (1999). *Why do banks merge?* (Working Paper n. 361). Banca d' Italia Research Department.

Focarelli, D., & Pozzolo A. F. (2001b). *Where do banks expand abroad? An empirical analysis* (Working Paper). Banca d' Italia Research Department.

Hagendorff, J., Collins, M., & Keasey, K. (2008). Investor protection and the value effects of bank merger announcements in Europe and the US. *Journal of Banking & Finance, 32*(7), 1333–1348.

Hagendorff, J., Hernando, I., Nieto, M., & Wall, L. (2010). What do premiums paid for bank M&As reflect? The Case of the European Union. *SSRN Electronic Journal*. https://doi.org/10.2139/ssrn.1537585

Hagnäs, T., & Pynnonen, S. (2014). *Testing for cumulative abnormal returns in event studies with the rank test*. https://doi.org/10.2139/ssrn.2479228

Hernando, I., Nieto, M. J., & Wall, L. D. (2008). *Determinants of domestic and cross-border bank acquisitions in the European Union* (Paolo Baffi Centre Research Paper n. 2008-33; Banco de Espana Working Paper n. 0823).

John, K., Mateti, R. S., Shangguan, Z., & Vasudevan, G. (2013). Does overvaluation of bidder stock drive acquisitions? The case of public and private targets. *International Journal of Banking, Accounting and Finance, 5*(1/2), 188–204.

Kyriazopoulos, G. (2016). Wealth effects from banks mergers and acquisitions in Eastern Europe. *International Journal of Economics and Financial Issues, 6*, 588–595.

Kyriazopoulos, G., & Drymbetas, E. (2015). Long-term performance of acquirers involved in domestic bank M&As in Europe. *International Journal of Financial Research, 3*(1), 100–116.

Leledakis, G. N., & Pyrgiotakis, E. G. (2016). *U.S. bank M&As in the post-Dodd-Frank Act era: Do they create value*? MPRA Paper 73290, University Library of Munich.

Leledakis, G. N., & Pyrgiotakis, E. G. (2019). Market concentration and bank M&As: Evidence from the European sovereign debt crisis. https://doi.org/10.2139/ssrn.3443171

Lepetit, L., Patry, S., & Rous, P. (2004). Diversification versus specialization: An event study of M&As in the European banking industry. *Applied Financial Economics, 14*(9), 663–669.

Pasiouras, F., Tanna, S. K., & Gaganis, C. (2007). *What drives acquisitions in the EU banking industry? The role of bank regulation and supervision framework, bank specific and market specific factors*. (Economics, Finance and Accounting Applied Research Working Paper Series, n. 2007-3). Coventry University.

Patell, J. M. (1976). Corporate forecasts of earnings per share and stock price behavior: Empirical tests. *Journal of Accounting Research, 14*, 246–276.

Rao-Nicholson, R., & Salaber, J. (2015). Impact of the financial crisis on banking acquisitions: A look at shareholder wealth. *International Journal of the Economics of Business, 22*(1), 87–117.

Rhodes-Kropf, M., Robinson, D. T., & Viswanathan, S. (2005). Valuation waves and merger activity: The empirical evidence. *Journal of Financial Economics*, *77*(3), 561–603. https://doi.org/10.1016/j.jfineco.2004.06.015

Shleifer, A., & Vishny, R. W. (2003). Stock market driven acquisitions. *Journal of Financial Economics*, *70*(3), 295–311. https://doi.org/10.1016/s0304-405x(03)00211-3

Tourani Rad, A., & Van Beek, L. (1999). Market valuation of European bank mergers. *European Management Journal*, *17*(5), 532–540.

Index

P. Wachtel et al. (eds.), *Creating Value and Improving Financial Performance*, Palgrave Macmillan Studies in Banking and Financial Institutions, https://doi.org/10.1007/978-3-031-24876-4

The manufacturer's authorised representative in the EU is Springer Nature Customer Service Centre GmbH, Europaplatz 3, 69115 Heidelberg, Germany. If you have any concerns regarding our products, please contact ProductSafety@springernature.com

Printed and bound by CPI Group (UK) Ltd, Croydon, CR0 4YY
28/09/2026
02224074-0005